THE VIRTUAL

This book looks at the origins and the many contemporary meanings of the virtual. Rob Shields shows how the construction of virtual worlds has a long history. He examines the many forms of faith and hysteria that have surrounded computer technologies in recent years. Moving beyond the technologies themselves he shows how the virtual plays a role in our daily lives at every level. The virtual is also an essential concept needed to manage innovation and risk. It is real but not actual, ideal but not abstract. The virtual, he argues, has become one of the key organizing principles of contemporary society in the public realms of politics, business and consumption as well as in our private lives.

Rob Shields is Professor of Sociology and Anthropology at Carleton University, Ottawa, Canada.

KEY IDEAS

SERIES EDITOR: PETER HAMILTON, THE OPEN UNIVERSITY, MILTON KEYNES

Designed to complement the successful *Key Sociologists*, this series covers the main concepts, issues, debates, and controversies in sociology and the social sciences. The series aims to provide authoritative essays on central topics of social science, such as community, power, work, sexuality, inequality, benefits and ideology, class, family, etc. Books adopt a strong individual 'line' constituting original essays rather than literary surveys, and for lively and original treatments of their subject matter. The books will be useful to students and teachers of sociology, political science, economics, psychology, philosophy, and geography.

Class
STEPHEN EDGELL

Community
GERARD DELANTY

Consumption
ROBERT BOCOCK

Citizenship
KEITH FAULKS

Culture
CHRIS JENKS

Globalization – second edition
MALCOLM WATERS

Lifestyle
DAVID CHANEY

Mass Media
PIERRE SORLIN

Moral Panics
KENNETH THOMPSON

Postmodernity
BARRY SMART

Racism
ROBERT MILES

Risk
DEBORAH LUPTON

Sexuality
JEFFREY WEEKS

Social Capital
JOHN FIELD

The Virtual
ROB SHIELDS

THE VIRTUAL

Rob Shields

LONDON AND NEW YORK

First published 2003
by Routledge
11 New Fetter Lane, London EC4P 4EE

Simultaneously published in the USA and Canada
by Routledge
29 West 35th Street, New York, NY 10001

Routledge is an imprint of the Taylor & Francis Group

Typeset in Garamond and Scala
by Keystroke, Jacaranda Lodge, Wolverhampton
Printed and bound in Great Britain by
Biddles Ltd, Guildford and King's Lynn

British Library Cataloguing in Publication Data
A catalogue record for this book is available from the British Library

Library of Congress Cataloging in Publication Data
Shields, Rob, 1961–
The virtual / Rob Shields.
p. cm. — (Key ideas)
Includes bibliographical references and index.
ISBN 0–415–28180–6 (hb) — ISBN 0–415–28181–4 (pb)
1. Information society. 2. Virtual reality—Social aspects.
I. Title. II. Series.

HM851 .S546 2003
303.48′33—dc21 2002027534

ISBN 0–415–28180–6 (hbk)
ISBN 0–415–28181–4 (pbk)

For Sophie,
who was virtual

Contents

Illustrations

FIGURE

TABLES

ACKNOWLEDGEMENTS

Books may be written, even researched, by individuals but they condense the sentiment and energy of many people. Books sometimes rise out of otherwise unglimpsed communities, and rise above the babble of more obvious contexts such as local groups or universities. Although they may never have met, books reveal communities of spirit. Few pay close attention to Acknowledgements. No one teaches their art. But, they hint at the multi-coloured trajectories from virtual and abstract into the concrete. These tracings of obligation and indebtedness are signs of social bonds and attachments otherwise denied a place in scientific texts. They remind us that research and writing is not merely a matter of 'schools' but of friendships, of clans and of communities. Acknowledgements are the places where love and gratitude are frankly confessed in academic texts.

This book takes stock of and draws a line beneath a decade of discussions of various forms of virtuality. The number of books – popular and learned – published on digital virtuality in the late 1990s is staggering. I owe a debt to these precedents, however much the public debate was conducted in terms that might in hindsight be thought muddled. Students and colleagues in a number of colloquia have helped clarify my thoughts. The Summer School of the Department of Geography at the University of Oulu, a Varenius seminar organized by the US National Center for Geographic Information and Analysis at the University of California, Santa Barbara and follow-up research on Internet communities in Ottawa funded by the Center and the US National Science Foundation, the 'Performing

Virtualities' graduate student symposium and 'Virtual Society? Get Real!' conference organized by colleagues and students at Brunel University under the auspices of the ESRC 'Virtual Society?' programme, and the colloquium on 'Presence and Absence: Fluid Networks' organized by John Law and sponsored by Nokia were outstanding examples of intellectual community. This book is one result of a generous Social Science and Humanities Council of Canada Strategic Grant under their 'Challenges and Opportunities of the Knowledge Based Economy' theme (Further information: http://www.carleton.ca/kbe).

Many others' comments and asides were more carefully measured than they would have guessed at the time. MacArthur and Elizabeth Shields and Bohdana Dutka were invaluable. Zoe Sujon and Jane Hampson kept this and many other projects moving when they would otherwise have been shelved by day-to-day concerns. I prized their comments and was lucky to be inspired by superb graduate students at Carleton University, including Anne Galloway, Walter Henry, Zoe Sujon and others in a seminar on New Media and Virtual Spaces, Adam Fiser, Heather Bromberg, Derek Foster, and others in seminars on cultural theory. Edwina Taborsky, Dan O'Connor, Suzan Ilcan, Petra Halkes, Joost Van Loon, Greg Elmer, Ian Roderick, Nicholas Packwood, Penny Ironstone-Catterall, Ken Hillis, Kevin Hetherington and many others gathered around the warmth of conversation and debate in the journal *Space and Culture*. Mari Shulaw, Ann King and the editorial staff at Routledge balanced patience and perseverance to get the book published. Finally, the patience of my family, the extended Dutka family, Bohdana and Sophie never wavered. They provided sustaining love, support and inspiration, virtual and concrete, over the entire course of the project.

Introduction

The virtual has become a key organizing idea for government policies, everyday practices and business strategies. What do we mean when we describe something as being 'virtual' – such as 'virtual space' or 'virtual team'? This book aims to help you better understand the virtual and why this category of things is suddenly important in business, government and in everyday life.

Today, the word 'virtual' is often used as a proper noun – 'The Virtual' – a place, a space, a whole world of graphical objects and animated personae which populate fictional, ritual and digital domains as representatives of actual persons and things. Commentators have not failed to remark that these avatars, video-game characters, software agents and virtual objects not only stand in for flesh-and-blood persons and physical places but they can have significant and shocking impacts on the real-life status and well-being of people. However, the more mundane case of virtuality includes lines of code in a database which record and police a person's financial transactions and debts. This 'credit profile' is one's virtual identity for transaction purposes as far as banks and merchants are concerned.

The chapters that follow examine the origins and meanings of 'the virtual' as a concept and what it means for people in everyday life under global capitalism. Beyond merely defining and mapping the spreading popularity of 'the virtual' as an idea, this book is a contribution to intellectual debates on the implications of a shifting relationship between the virtually real, and the material, the here-and-now world of the actually real. Cases of the virtual will be discussed in relation to three main categories: it will be contrasted with the 'concrete' and related to 'abstractions' and 'the probable'. The virtual may be found in ritual, religious debate, in architecture and art. The digital virtuality of the global Internet, simulations and virtual reality is only the latest incarnation of the virtual.

Examples from history show that an understanding of the virtual was commonplace. However, the historical importance of today's shift may be found in the rising popular faith in intangible essences, a focus on popular opinion, perception and insecurities as well as on tangible dangers or probable risks. Governments face a dilemma in that policies cannot be created to do more than assuage a sense of insecurity; businesses face the challenge of branding their products as much as in delivering actual quality or service. Both must balance the virtual and the concrete, but many fail, as case studies of Enron, terrorist attacks, pollution scandals and telecommuting will demonstrate.

We do not face a digital virtual utopia, and it is likely that fears over a 'digital divide' in access to the Internet will be judged in hindsight to be part of a campaign to boost technology. But the virtual raises profound issues regarding our attitudes and actions towards risk and our understanding of the importance of balancing the virtual with the concrete (in economics and everyday life), and the virtual and the abstract (in our culture and values).

1

THE RETURN OF THE VIRTUAL

> A whole new lexicon has arisen that seeks to capture the new ways of working . . . including 'Web enterprises', 'virtual organisations', 'virtual teams', 'teleworking' and so on.
>
> (Jackson, 1999: 3)

Do you think that there is anything new about the virtual? If so, you will be surprised to learn that in 1556 Thomas Cranmer was executed in large part because of his affirmation of the virtuality of the Eucharist. Similar charges were levelled against the reformation theologians Luther and Zwingli. Indeed, debates surrounding the virtual and practices of virtuality have a long history. This chapter introduces the historical importance and associations of the virtual as an aspect of cultures in Europe and other parts of the world. Sections introduce historical virtualities and develop the argument for the historicity of the virtual, as follows:

- Key definitions of the virtual include not only the virtual as essence or the 'essentially so' but the notion of 'virtue'.
- Virtual spaces and understandings of virtuality have a long history in the form of rituals, and in the built form of architectural fantasies and environments.
- Examples include: Christian reformation debates on the virtual in the Eucharist; baroque *trompe-l'œil* simulations and virtualities; liminal zones and rituals.
- Virtualism is the late twentieth-century fad for computer-mediated, digital virtuality, which draws on and repeats the historical forms of the virtual.
- However, it afforded a utopian moment despite the manifest contradictions of consumer hype and technological optimism.

DEFINITIONS OF THE VIRTUAL

> *The virtual*: Anything, 'that is so in essence or effect, although not formally or actually; admitting of being called by the name so far as the effect or result is concerned'.
>
> (*Oxford English Dictionary*)

Dictionaries define the virtual in everyday life as 'that which is so in essence but not actually so'. Thus we speak of tasks which are 'virtually complete'. More philosophically, the virtual captures the nature of activities and objects which exist but are not tangible, not 'concrete'. *The virtual is real but not concrete*, as we will be arguing in Chapter 2. Dreams, memories and the past are famously defined by Marcel Proust in his correspondence on *Remembrance of Time Past* as virtual: 'real without being actual, ideal without being abstract.' Proust's comment provides an important historical model for the use of the term today.

The noun 'virtual' comes to us from the Latin *virtus*, meaning strength or power. By the medieval period *virtus* had become *virtualis* and was understood in the manner we might understand the word 'virtue' today. In this older usage, a 'virtual person' is what we might understand in more contemporary usage as a person of some outstanding quality:

> '*Virtual*: Latin 1. *virtus* 2. *virtuosus*. Possessed of certain physical virtues or capacities; effective in respect of inherent natural qualities or powers capable of exerting influence by means of such qualities (rare)'.
>
> (*Oxford English Dictionary*)

The related term, 'virtue', is a personal quality, 'The power or operative influence inherent in a supernatural or divine being' (*OED*). Virtue is 'an embodiment of such power' (*OED*). In the less celestial terms of ethics, virtue is the 'conformity of a life and conduct with the principles of morality; voluntary observance of the recognized moral laws or standards of right conduct; abstention on moral grounds from any form of wrong-doing or vice' (*OED*). Virtue is also 'chastity, sexual purity and industry, diligence', or 'personified moral quality' (*OED*). Examples of this usage trace back to 1398. As an adjective, a 'virtual person' was what we might today call a morally virtuous or good person: a person whose *actual* existence reflected or testified to a moral and ethical *ideal*. Virtue was the power to produce results, to have an effect. Some even argue that 'the virtue of something is its "capacity" or efficacy' (Haraway, 1992: 325). But *Virtu* is more an open, creative potentiality.

Today, 'the virtual' is still redolent of its barely masked links to the concept of *virtue* (with which it shares a root in the medieval Latin *virtus* – from *vir*, 'man'). Few remember that an order of angels was said to be called 'The Virtues'. However, women's chastity is still mentioned in dictionary definitions of 'virtue', a difficult matter to verify empirically, which has long been the essence of patriarchal preoccupations. This strange twist in definitions in which we have ended up at 'chastity' points to the mixture of ambiguity and high stakes in social definitions of the virtual:

> no matter how big the effects of the virtual are, they seem somehow to lack a proper ontology. Angels, manly valor, and womens' (*sic*) chastity certainly constitute, at best, a virtual image . . . the virtual is precisely not the real; that's why 'post-moderns' like 'virtual reality.' It seems transgressive.
>
> (Haraway, 1992: 325)

VIRTUALISMS IN HISTORY

The virtual certainly has been controversial in the past. Where today's users of virtual reality or members of online virtual teams complain of carpal tunnel syndrome, in earlier epochs other notions of the virtual could carry the punishment of death. The argument here is that the virtual has long been significant as a cultural category, as part of the human mental toolkit. Furthermore, two brief examples suggest that we could learn a great deal about the social actualizations of the virtual from historical cases. The virtual has long existed in the form of rituals, and in the built form of architectural fantasies and environments.

In fact, if the virtual has meanings of 'virtue', of being 'almost-so' or 'almost-there', one does not need to look far to find virtual worlds which surround us or their historical counterparts. Virtual worlds are simulations. Like a map, they usually start out as reproducing actual worlds, real bodies and situations; but, like simulations (see following section and Chapter 2), they end up taking on a life of their own. Somewhere along the way they begin to diverge, either when it is realized that no map can be so complete that it represents an actual landscape fully, or when they become prized as more perfect than messy materiality. As virtual worlds, they become 'virtuous', utopian. Virtual worlds become important when they diverge from the actual, or when the actual is ignored in favour of the virtual – at which point they are 'more real than real', as Jean Baudrillard, a theorist of the ironies of late twentieth-century cultures, has pointed out. An example is found in the way representations of the health of stock-markets, as expressed in, say, the charts and econometrics of a computerized news service, routinely stand in for the actuality of the economic life of nations half a world away. This 'hyper-real' quality implies that the virtual has to be taken into account on its own terms, because it is no longer simply a reflection of the actual (see Chapter 7).

Historical impacts of 'the virtual': the Reformation

Rather than a matter of angels or other virtual beings, the debate concerned the mystical transubstantiation at the centre of the Christian Eucharist – the conversion of bread and wine into the body and blood of Christ. Actually real, material body and blood, insisted the Church. 'Virtually real', argued Reformation theologians.

In October 1517 Martin Luther nailed his Ninety-five Theses to the door of the church in Wittenberg. At the heart of his objections was the catholic doctrine of the Real Presence of Christ in the Eucharist. Mass as a sacrifice or as a good work which could be charged for was anathema to Martin Luther and one of the key errors to which he objected (Luther, 1523: 441, 32n). Reformers viewed theories such as transubstantiation as an unnecessary detour to explain the miracle of the 'Real Presence' of Christ at each and every re-enactment of the Last Supper in rational terms, when any miracle by definition defies any such explanation. The substance of the Eucharist 'is, and remains, bread' (Luther, cited in Brooks, 1992: 20; see 1 *Corinthians* 10.16). Accordingly, the faithful need only believe.

As Protestantism spread, controversy arose over the status of the Eucharist. One famous trial for heresy took place in September 1555. Archbishop Thomas Cranmer was examined for heresy in the Church of St Mary at Oxford. Seated 'in the East end of the said church, at the high altar', on a chair set on a 'solemn scaffold . . . ten foot high . . . under the sacrament of the altar' (Cranmer, 1846: 212; cf. Foxe, 1877, VIII: 44, cited in Brooks, 1992) the Archbishop was cross-examined on his teachings regarding the reality or virtuality of the Eucharist. Orthodox Catholics held that it was 'necessary to be believed as an article of faith, that there is the very corporal presence of Christ within the host and sacrament' (Cranmer, 1846: 246). 'Transubstantiation' as a belief and doctrine had its origins in the theology of St Thomas Aquinas. In each and every Mass, Christ was present. In each and every Mass, a sacrifice took place.

The beginnings of the Anglican tradition lie in Cranmer's attempt to tread a fine line between the Protestant influence of Martin Luther and Zwingli and his own convictions that the truth of the Eucharist be judged independently, empirically and with 'discrimination' (Robinson, 1846–1847: 13). But persuaded by dissenting preachers, this stout defender of Catholicism came to agree that 'the Scripture knew no such term of "transubstantiation"' (Foxe, 1877, V: 501). 'Transubstantiation' was the transformation of mundane bread or a host into a piece of the body of Christ. The essence of the debate was the question of whether this occurred literally and superstitiously. The Calvinists espoused a doctrine of 'Virtualism' – of Christ's virtual presence in the Eucharist. Cranmer's understanding gradually changed away from a belief in the Real Presence of Christ in the bread and wine towards a position favouring the symbolic and virtual presence of Christ in the Eucharist.

Although the result was actually disastrous for the Archbishop, a hundred years later in 1654 a source cited by the *Oxford English Dictionary* could publicly proclaim: 'We affirm that Christ is really taken by faith . . . [although] they say he is taken by the mouth and that the spiritual and the virtual taking him . . . is not sufficient.'

The doctrine of virtualism raised questions concerning the way we understand presence – must it be concrete and embodied or was 'essentially present' good enough? Was there anything there if it was virtual? The same questions are raised today concerning online environments and virtual reality, and are treated in the chapters that follow. Are they real? Should they be given the same regard and dignity as other spaces of interaction?

Baroque cyberspaces

One of the most interesting historical uses of the virtual anticipates the way in which people now refer to virtual realities or virtual teams. This is found in the discussion of mirror reflections as 'virtual images' and of the way we experience dreams as 'virtually real'. In optics, a 'virtual image' is formed by the apparent, but not actual, convergence of light rays to make an apparent but not exact counterfeit of the real. This is not simply a matter of perfect resemblance, however, for the image is reversed left to right. The image is virtual in that it suggests a potential mirror-world on the other side of the glass, an early precursor of the power of simulation. Illusions, mirrors to extend the space of a room (such as the Palace of Versailles' Hall of Mirrors) and *trompe-l'œil* decoration fascinated eighteenth- and nineteenth-century writers.

If cyberspace is a 'consensual hallucination', in the words of the novelist who coined the term, William Gibson (1984: 67; see Chapters 3 and 4, this volume), then cave paintings might well count also. But skipping backward only 200 years, and much closer to our time, another historical moment celebrated the virtual to produce the first elaborate virtual environments – often in the form of the interior decoration of churches. This indulgence in *trompe-l'œil* contrasted with the dislocation and wars of eighteenth-century Europe, the first state powers asserting a harmonious, ecstatic world, in part as an expression of their power. These simulations were made to appear to defy gravity.

> The heady lure of these mystical works is based on their elaborate continuities of human and fictive space. . . . They pair techniques involving the creation of a dreamscape, and the provision of [human] figures for identification that call the viewer to enter fictive space, changing with their movements, inviting their co-authorship. They are fundamentally navigable . . . 'spaces of persuasion'.
>
> (Maravall, 1986: 74–5, quoted in Cubitt, 1998: 75)

Baroque architecture and decoration rendered a dramatic space of swirling movement beyond the cares of the subluminary world into paint, plaster and marble. Not only did painted scenes of heavenly delights on vaulted ceilings trick the eye; the buildings were celebrations of forced perspective both in their floor plans and sections. Dominant lines of cornices, and rows of columns were shifted off of a right-angled grid to converge slightly, giving an impression of grandeur and distance.

> At its pinnacle, the Baroque offered the thoroughly mediated interactivity of audience participation in the spectacle of its own rule. . . . [It] 'was, like postmodernism today, at once a technique of power of a dominant class in a period of reaction and figuration of the limits of that power' . . . we need to understand the culture of spectacle in the first Baroque as the beginnings of our own. To understand that the vertigo of imperial expansion, the terrors of absolute power and the morbid fascination with decay and mortality have been transformed into these virtual architectures is to catch a glimpse of the emergence of our own obsessions with the universe as our object of possession, our anxieties about absolute commodification.
>
> (Beverley, 1993: 64, quoted in Cubitt, 1998: 75)

Virtual environments have been less spectacular in their treatment of space due to technical limitations. However, they share the concern of the baroque church ceiling to draw the viewer into a spectacle which transcends the everyday spaces of the temporal world, at the same time pushing that participant away as a 'fallen' mortal. The mind and soul could escape, but in both cases the body is a dead weight which pulls one back to Earth. Angels indeed – these spaces solicited a separation of the mind and body into a virtual and concrete pair: the soul and the flesh. For the former, salvation came through the powers of the state and its church; for the latter, abjection and domination as a 'bare life' (cf. Agamben, 1998) worthy not of lofty institutions such as the state but of the soil.

Some of the first commercial immersive environments, such as nineteenth-century panoramas, drew huge paying crowds to see the world as controlled spectacle. Like a diorama in a museum which has been constructed and arranged to show the ecology in which an animal lives, panoramas attempted to create a virtual environment via a 360-degree painting viewed from a central viewing platform. Into these circular paintings 'it was possible to project yourself imaginatively, exploring the *mise-en-scène*' visually, as earlier Europeans had marvelled in the baroque ceilings of their basilicas. In

> a curious inversion of the panopticon, placing the subject in the centre of the field of vision, radiating out into a world prepared for ocular discovery, placing . . . the power of universal vision firmly in the eye of the mass spectator, a bizarre democratization of the aristocratic gaze, first as panoptic professional, and then as the world-spanning, mobilized look of the sovereign individual' – the paying spectator.
>
> (Cubitt, 1998: 78–79)

Panoramas

The gazebo-like central viewing platforms of famous circular panoramas such as, for example, the *Mesdag Panorama* allowed viewers to look out on a circular painting mounted in a rotunda. The 14m-high *Mesdag Panorama* presents a seaside scene. This virtual beach has been on display since 1881 in The Hague, Netherlands. It allows one a vicarious view of a timeless, harmonious and cultured nature. Patrons would ascend from a staircase below into a viewing platform constructed like a gazebo which blocked out the ceiling of the building and prevented the viewer from getting too close to the painting (Halkes, 1999: 84). This is not an interactive environment – nothing in the painting changes to respond to the viewer, nor was the scene ever peopled by actors to enhance the illusion. It is not simply a representation but a simulation in which real sand conceals the bottom edge of the painted beach scene (Halkes, 1999). The panoramas were extravagant attempts to not only mimic reality but to outdo actual experience (in this case of the popular seaside destination of Mesdag), by relocating the viewer to a panoptic and omniscient position.

As Crary argues in his book *Techniques of the Observer*, vision had been understood as the privileged sense of truth and of divine revelation. Seeing was believing. The pinhole camera obscura was the icon of classical vision because it revealed the physics of light and images. By contrast, he argues, the panopticon and stereoscope broke with this timeless model. These are the icons of the embodied, binocular vision of the nineteenth century. Unaided vision was shown to be all too human. It depended neither on revelation nor on laws of optics but on physiology and the imperfect, ageing biology of the human eye. The inverted images seen through pinhole cameras or in a camera obscura demonstrated this in physics and optics (Halkes, 1999). The stereoscope (an apparatus for 3D viewing by combining two photographs of the same scene, one slightly displaced from the other) and zoetrope (in which a series of drawings of an action, spinning on a circular tape or shade, were viewed through a slit, giving a cinematic appearance of moving images) depended on human binocular vision to make sense of otherwise nonsensical images (Crary, 1992: 67ff.).

> the pictorial panorama was in one respect an apparatus for teaching and glorifying the bourgeois view of the world; it served both as an instrument for liberating human vision and for limiting and 'imprisoning' it anew. As such it represents the first true visual 'mass medium'.
>
> (Oettermann, 1997: 7; see also Halkes, 2001: 60)

But more recently, Halkes incisively argues that the panoramas did not break with classical vision once and for all, nor did they testify to an alienation from an all-embracing truth. The panopticon was not part of a linear evolution of the sense of vision.

Rather the panopticon was an example of more complex desires in the nineteenth century for a classical vantage point analogous to the eye of God – even as it was being displaced at the time. Advances in technology and medical understanding removed vision from the order of divine revelation while visible appearance was displaced in favour of the microscopic and invisible in the sciences (see Mizroeff, 1999; Friedberg, 1993).

In some ways, today we are back to the panopticon. The rise of digitally simulated objects and environments raises similar issues. They displace unaided vision and the frail bodies as the standard of insight and performance. Digital simulations both liberate and incarcerate, displacing the original material world in favour of virtual environments (see Chapter 3).

Liminoid virtualities

Although these are only two examples, none of the many historical virtualities required the purchase of computers or online subscriptions. But we can clearly find historical types of virtual realities, fictions, simulations and perception games which tricked the mind and body into feeling transported elsewhere. Retrospectively, it is clear that there has been a history and succession of 'virtual worlds' which anticipate the ability of information and communications technologies to make present what is both absent and imaginary. The cinema is one example, but any number of rituals create, through a willing suspension of disbelief (for Euro-Americans), milieux in which rules other than those that govern the face-to-face interactions of actual bodies are the norm (for example, flashbacks and other temporal reorderings, leaps from scene to scene and 'superhuman' powers).

For most cultures, however, collective 'conjuring' of altered modes of perception and understanding are more common practices. These virtual spaces that populate the anthropological literature are lived more strongly than the mere 'consensual hallucination' envisioned for cyberspace (cf. Gibson, 1984). Rituals inaugurate liminal zones which are the performative settings for rites of passage such as puberty or marriage (Turner, 1974). These zones allow what is often a symbolic death or removal from one social status and birth into another. Initiates first lose their status and, after undergoing the appropriate rituals, are received back into the society and the space of the everyday with a new status. In between is a 'time out of time' on the '*limen*' (threshold) of membership or a new status. In this space, initiates are instructed in their new identity and responsibilities. The bride and groom's walk down the aisle at a wedding is a common example familiar in European and American societies. The wedding service is a liminal time and space. In it, the bride and groom enter according to strict customs. Harking back to ancient patriarchal traditions, the bride is escorted down a central aisle and 'given away' by her father or another representative of her family. The couple receive instruction from the priest and promise to care for each other, even if in what seems like code today – 'I promise . . . to have and to hold' and so on. The bride and groom then exit down the aisle as a new, socially recognized couple.

Like Janus, the double-faced god of doorways and portals, the border between the everyday and sacred, ritual spaces face both inward and outward, creating an equivocal, ambiguous zone – a zone is not just a line, but a strongly marked, interstitial space. '*Limen*' are thus 'threshold' spaces in which one is neither 'in' nor 'out' (Turner, 1974). A key part of the transformation is the suspension of everyday social norms to allow a rearrangement of the social order, conferring new status and allowing society to acknowledge and recognize the new identity of those who have been the focus of the ritual. As such, liminality offers a utopian moment in which the weight of limiting social regulations is lifted. Liminality is crucial to the adaptive powers of a culture.

Liminal zones are virtual environments or spaces. The bride and groom remain quite close by; they do not literally and materially travel from one place to another. The rules of quotidian face-to-face life are suspended or even inverted in a carnivalesque of norms. In their place, special rules of engagement rule the moment and the space. Victor Turner's famous dictum states that liminality is 'betwixt and between' stages in the life process, located between the urban/civilized/members and the wilderness/nature/outsiders (Turner, 1974). Of less life-changing status, there are many examples of *liminoid* spaces and genres in any society – the Web, vacation resorts, theme park environments not to mention specific holidays and events (Shields, 1989). In contemporary society, liminality has been stripped of its transformative power to become a commodified experience, and no more so than in the tourism and leisure industries (Shields, 1991) and online (Shields, 1996; Silver, 2000).

Like liminal zones and events, virtual spaces are 'liminoid' in that they are participated in on a temporary basis, and distinguished from some notion of commonplace 'everyday life'.[1] Virtual space is not only betwixt and between geographical places in a non-place space of telemediated data networks, but participants take on specific 'usernames' or identities, and many surreptitiously engage in activities they might not otherwise consider. Computer-mediated, digital forms of virtuality are continuations of long-running processes; to be understood they need to be linked back to a history of cultural forms such as the liminal.

However, is this loss of the liminal a degradation of the virtual in digital virtual spaces? The technology and fixed programming code of virtual realities supercharges and often overpowers the qualities of liminality. The greatest power of digital virtuality – and perhaps its most widely discussed feature – has been in providing a matrix in which new modes of being and practices of becoming could be experimented with. In its early stages through the 1970s and 1980s, few and tenuous guidelines were provided for *metaxis*, the leap from the concrete to the virtual. This was usually a leap of imagination but in the case of online gaming it became merely a question of adjusting a computer interface (see Chapter 2). Metaxis is the key conceptual sleight of hand in allowing users to imagine leaving behind identities in one realm to become something/someone else or to play an entirely different role (for example, in a role-playing game). The charged, affectual space of online games and chats gained its character as an extension of the rhythms and encounters of virtual bodies, sociable exchanges and animated tracings of hypertext links, none of which the space pre-existed except abstractly. A liminal zone provides the potential for assuming new identities, and thus the virtual became a liminoid space; not one directed at rites of passage, but rather at experimentation – like that other, sacred liminoid space of advanced economies, the scientific laboratory.

The virtual rebounds on the material and the abstract, changing the Enlightenment tradition of simple dualisms not only of here and there, inside and outside, but of concrete and abstract, ideal and actual, real and fake, transcendent and immanent. The either–or model is shifted in a tangible and everyday manner into a system of hybrids of the old dualisms which are best understood as intensities and flows (see Shields, 1997). The virtual infects the actual as a metaphor which has moved from the realm of digital domains and computer technologies to become an organizing idea for government policies, everyday practices and managerial strategies. The virtual shifts the commonsense notions of the real away from the material. The virtual, as in a 'virtual organization', is more heavily invested with notions of collective performance and *inhabitation* than a priori architectural objects such as 'the factory' or 'the office'.

Like other liminoid zones under capitalism, such experiences and sites generally become commodified as package tourist attractions, not sacred places which are the sites of cures or pilgrimage destinations. From the virtual as a threshold to the effervescence of cultural margins, the Internet becomes more and more a pay-per-view, pre-screened information service. Much of the popular discussion of computer-mediated communications amounts to domesticating virtual spaces and bringing it out of its liminoid status – a realm of illicit information (how to build a nuclear bomb and so on), the resort of the repressed that contemporary culture generally excludes or refuses to grant a place to (the obese, those physically challenged in one way or another), an arena in which forbidden desires are unleashed, and a subculture populated by mythified figures such as the hacker.

UTOPIAN VIRTUALISM

The hype around digital virtuality over the past decade has been more about myth and less about actual cyberspaces. As a fad and myth, virtualism is itself virtual. Symptoms of virtualism include exaggerated expectations of anything described as 'virtual', and unrealistic expectations that digital technologies will solve social problems. The boom in technology stocks and enthusiasm for virtual reality hinted at the ongoing expectations of the virtual. In line with its historical definitions, it carries a certain promise of positive potential or virtue. Portrayed as enabling a human virtuosity beyond the limits of the body or gravity, the legacy of the baroque echos through the claims of Silicon Valley entrepreneurs.

The explosion of virtual reality as well as more mundane virtual spaces is that it allowed a utopian moment of gaiety that was arguably the most significant Western, and even more specifically American, counter-cultural moment since the 1960s. Although it was reabsorbed into the commercial mainstream, its utopian and liminal moments commodified and packaged into experiences for sale or vague promises of excitement attached to the purchase of a home computer, virtualism marks the culture of the close of the twentieth century as surely as stock-market booms marked the economy.

Unlike the 1960s this moment of cultural effervescence and optimism was not limited purely to one demographic group such as the young or the wealthy but was participated in by a range of consumers and producers who stretched from the young inventors of video-games (in their early teens) to financiers and investors who supported and 'bought in', socially and psychically, to the utopian dreams of, first, Silicon Valley entrepreneurs and, later, dot-comers. Statistics showed that older people, poor households and young black men neither dived into the consumer frenzy for technology (the devices themselves) nor acquired the skills to enter and keep abreast of the rapidly evolving industry. Computers came to appear as essential, as a necessity. Despite all this hand-wringing, the last two decades of the twentieth century witnessed an explosion of utopianism into the mainstream which has been only partially quelled by the familiar journalistic doubt, accusations of political naivety and an unwarranted faith in technology to transform social relations and redress inequalities from a personal to a global scale.

A remarkable element of this process was how quickly the paradigms on which information and computing technologies were based evolved and matured. Short, two-year cycles of novelty followed by obsolescence which had been finally rejected by automobile consumers reappeared in the computer industry. Software and machinery that did not work or was so insecure as to be a dangerous liability together with inflated promises and hype echoed some of the cars of the 1950s and 1960s. The car, after all, was the greatest vehicle not only of people and materials moving from place to place, but of myths and dreams – virtual delights and transports. However, as Poster notes,

> the history of electronic communication is less the evolution of technical efficiencies in communication than a series of arenas for negotiating issues crucial to the conduct of social life; among them, who is inside and outside, who may speak, who may not, and who has authority and may be believed.
>
> (Poster, 1990: 5)

SUMMARY

This chapter has considered the virtual as defined in dictionaries and encountered in historical forms of social interaction. The basic dictionary definition of the virtual is 'anything that is so in essence . . . although not . . . actually' (*OED*) as in a task which is 'virtually complete'. A related term, 'virtue' suggests the intangible or latent quality of virtuality – there but not necessarily obvious to the senses. Historical virtualisms abound in simulations and representations that take on a life of their own (such as Baroque church interiors and Panoramas). But religious debates over the nature of virtual presences, such as during the Reformation, have been ugly in the past. The close affiliation between the virtual and liminality is especially significant for cultural and anthropological analyses of the 'prehistory' of contemporary European and American fascination with digital virtualities. In its use to conjure altered perceptions and understandings, the virtual overlaps with liminal rituals such as rites of passage. Liminal zones are social spaces in which initiates are 'betwixt and between' old and new social statuses and identities. Today's commercialized, digital virtualities are liminoid in that they derive from the liminal but do not entail rites of passage. The utopian tint and optimistic outlook of late twentieth-century virtualism indicates its positive potential across social groups. However, later chapters will consider the exclusive quality of digital virtuality. While suspicious of a sales pitch that mobilizes fears of a 'digital divide', the question of who speaks and who gains entry to digital virtual environments and simulations is an important one. More profoundly, the implications of the virtual for our attitudes and actions towards risk and our understanding of the importance of balancing the virtual with the concrete (in economics) and the virtual and the abstract (in culture) will be probed in the chapters that follow.

2

THE VIRTUAL AND THE REAL

What is 'the virtual'? The virtual calls into question our preconceptions about the actual, demanding that we broaden our understanding of reality. Beginning with the basic meanings of the virtual and its contrast with the actually real, this and the following chapters examine:

- The cultural impact of computerization as a new digital virtuality.
- The significance of the virtual in leisure time, family life and for simulation and video-gaming subcultures.
- Workers' experience of and roles in the virtual workplace.
- The virtualization of firms and organizations, including successes and failures.
- The morality and ethics of virtual social relationships at a distance over the Internet, including attempts at the moral regulation of the Web.
- The implications for everyday life off-line, including the experience of unwired societies and those excluded from the virtual worlds of computer-mediated telecommunications.

This chapter examines how the virtual is often contrasted with the 'real' in commonsensical language by many writers who have not paused to examine the implications of the terms they are using. Other commonsense cases include the way we talk of the 'virtually real', the 'virtually completed' task or the 'virtual team'; and the way in which we understand ritual, faith and our memories. We are interested in slippages in meaning, the way in which new understandings of the virtual are coming to prevail not only in professional and public cultures but in everyday life. This process occurs through myriad techniques, not only through digital communications.

As argued in Chapter 1, virtuality appears in various forms throughout history which are sometimes explicitly called virtual. The idea and word are by no means new. But today's tight connection of the virtual to digital hardware and software is a new form. It represents a return of 'the virtual' in our social activity. Some would just dismiss the term as an overused and underdefined label. However, this ironically recognizes that, at a minimum, 'the virtual' is one of the most important marketing terms for the high-tech sector which is claimed to drive the development of a putative high-tech, knowledge-oriented 'virtual society'.

Still, 'virtual' is often meant to signify an absence, unreality or non-existence. Everyday talk in the media equates the 'real' with concreteness, material embodiment, tangible presence and reliability. These definitions suggest that the virtual is a type of wooden nickel, not 'the real' thing, valueless and without dignity. So why is the term so widely used? Fortunately, popular wisdom is something different from talk and we routinely deploy the word 'virtual' as a place-holder for important forms of reality which are not tangible but are essential and necessary to our survival.

Beyond this sceptical stance, the popularity of the virtual as an adjective applied to almost everything points to barely acknowledged but widespread desires and beliefs. The multiple uses of the term 'virtual' hint at more than the digital: the term has connotations of effectiveness and success. 'Virtual' is a space; it is places, relationships, and implies values. To understand the term and the power of its associations is to be armed with a tool for cutting through hype to the lasting core of technological and economic change. I will argue that it is indicative of a sea change in cultural attitudes: we are becoming more comfortable with absence, more nuanced in our use of abstraction, and more dependent on the past as a bastion of identity in the face of a global cultural and environmental future which no one can predict.

VIRTUALLY REAL

As the discussion of definitions in Chapter 1 illustrate, the virtual is often defined in contrast with '*the real*'. However, this then raises the issue of what 'the real' is. For psychologists and physiologists, a physically real object is one that can be verified by others and its movements tracked by most firsthand observers who perceive it (cf. Shapiro, 1995). But when one transfers a computer image or file, can it be said to move in the same physical way? No. The virtual is neither absence nor an unrepresentable excess or lack.[1] The file moves and is conventionally verifiable for most computer users, and is 'real', so we need to break down the commonsensical notion 'reality' into more fine-grained concepts. Although few reflect on it, it turns out that this is something most people do anyway – we are far more sophisticated in our day-to-day manipulation of virtual and actual objects than we might suspect. Although this topic is worrying only for a few, perhaps not very sophisticated academics, one commentator has argued that this is the true value of the virtual – to directly confront the question, 'What is reality?' (Woolley, 1993). For example, virtual reality and simulation technologies (e.g. flight simulators, role-playing games and 3D architectural displays) attempt to replicate the sensory information of the physical world in order to present a constructed 'information-world'. While common sense appears to supply a ready answer to the differences between the virtually real and the actually real, the issue of 'the real' has generated centuries of philosophical debate. Entire fields of philosophy have developed around this question. Ontology (studying 'what exists?') and epistemology (studying 'how we can be certain' about what exists) have examined such questions from many sides and provide a wealth of insights into the many forms reality takes.[2]

The virtual troubles any simple negation because it introduces multiplicity into the otherwise fixed category of the real. As such the tangible, actually real phenomena cease to be the sole, hegemonic examples of 'reality'. Further, the logical identity of the real with these phenomena is broken apart, allowing us to begin to conceptualize processes such as becoming in terms of emergence and dialogism (cf. Bakhtin, 1981) rather than only as a dialectical as a negation of existing identities (Laclau, 1996: 20–46).

Operating with a simple notion of the tangible and the original as the one and only 'actually real' leads to a series of conundrums over anything produced from a model or in a series, such as in the case of mass production. The solution is not to debate the reality of the virtual, but to develop a more sophisticated theory of the real and the ways in which the *virtual* and the *concrete* are different really existing forms, how they are related to each other and to non-existing *abstractions* and *probabilities*. To do this, we want to build up, out of its shadings and partial uses, a model of what people understand by 'the virtual'. This will allow us a strategic insight into how commonsense notions of the world at large are changing, and how people's understandings of their powers and possibilities in that world are following suit, with the result that they act in ways which would be unexpected according to previous models of reality – one which left out or did not value the virtual. Perhaps this will help us to understand what we mean by 'reality' these days.

In everyday usage, 'the virtual' has many meanings. Something 'virtual' might be distant, it might be something invisible but important, or it might refer to informal arrangements or latent factors. Even on a strictly local scale, the idea of 'virtual teams' has become an influential organizing idea for competitive businesses. Virtual teams are not only groups of workers who communicate through computer email and so on, but all teams that are assembled to address particular types of problem, to respond to crises or to pursue very specific projects – springing into action with the lightness of electrons, and winding up their operations at the conclusion of a project. If these teams are fleeting they can be recalled back into existence, like a computer file redisplayed on a video screen (see Lipnack, 1997). They are virtual if only because they are neither face to face nor propinquitous (local); rather they are far-flung, temporary and latent. Their supporting infrastructure is a rented communications link and thus they leave few tangible traces other than email records and archived video-conference recordings. A search of the Internet reveals not only 'virtual worlds' but also virtual hospitals; florists; virtual tours and virtual tourists; many games (Virtual Pool); towns (e.g. Virtual Springfield Mass., or Santa Cruz Cal.); music, malls, virtual girlfriends (Bernadette.net in Australia has long been one of the most famous websites); an ancient Egyptian 'virtual temple' (thoroughly contemporary and accessed via an American server), and a virtual Jerusalem (which leaves one wondering about whether or not heaven could be described as 'virtual').

Upon close inspection, popular uses of the virtual make it clear that people understand this as intimately tied to the tangible and actually real. Anything 'virtually so' is very close to being really so. 'Virtual' covers all things that are 'almost so' – unfinished jobs which we none the less call 'virtually complete', a second-hand car which is 'virtually new' and so on. Etymologically, 'the virtual' is exactly this: it is what is so in essence but not in form. The '*actual*' contrasts with the essential, conceptual or '*ideal*' quality of these common notions of virtuality. The opposite of the virtual, however, is the *concrete*.

SLIPPAGE

It is important to distinguish our approach here from Plato's philosophy of forms, in which ideal types informed and animated the actual manifestations encountered in material reality. The ideal of trees was a required 'essence' hidden in any given tree. Where Plato argued that this was the ontological basis of reality, we are showing how humans have a cognitive ability to substitute 'what is so in essence' for actual things themselves. We understand that *x*, *y* and *z* stand for quantity to be substituted into a calculation rather than requiring the actual objects to be lined up and enumerated. We enjoy the lifelike as much as the living, and with collect representations of the far away and the past. All these are virtualities. Fiction, imagination, memory, engineering and mathematics depend on this cognitive ability, as do representations, conceptualizations and all ideations (all non-actuals of every sort). Abstraction and the fabrication of purely cognitive representations and signs are interlinked with the same capacities required for the virtual.[3] Actual materiality accounts for the gross mechanics of the natural world, but neither quantum mechanics, nor social science, nor studies of digital transactions could be complete without attending to the virtual as much as to the concrete.

Even though the virtual retains a quality of something that is 'almost so', it can quickly come to appear to have real substance in and of itself. To describe something as 'virtual' indicates that it is not strictly according to definition, as in a 'virtual office', which is to say not literally an 'office' as one might understand an office to be, but an office 'in effect'. This example illustrates how being 'not quite', say, an office can shade into being a new form of the office which necessitates a change in the definition of offices and possibly of office work. Raymond Williams, one of the founders of cultural studies, once pointed out how these are cases 'of a definition of quality which becomes, through real usage, based on certain assumptions, a description of the world' and a self-fulfilling prophecy, moving what was once only a perception into being a worldview (Williams, 1981: 68). The virtual is by no means the first case of this shift from a specific description to an essentialized, self-evident way of the world, something seen as 'the nature of things' or all-encompassing context such that everything comes to be seen to be, in one way or another, as having a virtual component as well as a material existence. Williams uses just such an analysis to examine the manner in which the sense of the 'natural', as an 'inherent and essential quality of any particular thing', became a cultural notion of *nature* – 'the essential construction of the world' (Williams, 1980: 68). A quality becomes reified, or turned into a thing itself. On the one hand, this could be dismissed as a category mistake; but on the other, it highlights the manner in which the virtual is closely bound up with the concrete – it does not make sense to locate the virtual outside of the 'real' but rather to make it part of it. General qualities are virtualities that 'really exist'. They can co-exist and co-define an 'actual' object or process in the material world, the manifestation of which depends on the context or situation in which it takes place.

By using old terms in new ways, there is a slippage of definitions and a transference which takes place between originals and new technological forms, simulations and objects. This has been called a process of '*seconding*' or 'trafficking' between the traditional or known and the new (Franklin *et al*., 2000: 22–23). Seconding may reinforce or replace the original. In *Global Nature, Global Culture* Franklin *et al*. give the example of a trademarked cosmetic 'Virtual Skin' from the company Prescriptives. 'Women want liquid skin in a bottle', says a representative. The ad copy reads, 'make-up priorities have changed – we want foundation to hide blemishes, disguise shadows and be imperceptible. It has to look, act and feel like skin.' Culture and nature, the artifice of 'Virtual Skin' and actual skin, 'mimic each other's qualities such that they can hardly be differentiated, while the difference between them is precisely what makes this seconding or substitution desirable' (Franklin *et al*., 2000: 25). 'Virtual Skin' is claimed to retain both the essence of actual skin adding the virtue of improving over the original. This transferability between the actual and the virtual, in which a quality becomes the essence of the matter, appears again and again in advertising and business notions of the virtual.

THE VIRTUAL IS REAL BUT NOT ACTUAL

Proust commented that memories are virtual: 'real without being actual, ideal without being abstract.' Dreams and vivid memories may be mistaken for experiences which one is actually living. We may awake from a dream that seems so real, so 'lived' that for a moment we confuse it with an actual experience. It may even inspire us to action– to achieve our dreams. While we may recognize the difference between actual and these imagined or recollected events, the richness and power of such experiences makes them important to us and highly valued in many cultures.

But the virtual is not only contrasted with the actual. It is different again from the abstract and from the probable or possible. As Stivale (1998), commenting on the work of the philosopher Gilles Deleuze, comments, the opposite of the really existing is the *possible*:

> The possible is never real, even though it may be actual; however, while the virtual may not be actual, it is none the less real. In other words, there are several contemporary (actual) possibilities of which some may be realized in the future; in contrast, virtualities are always real (in the past, in memory) and may become actualized in the present.
>
> (Hardt, 1993: 16)

The *possible* is that which does not really exist, but could to various extents. At one extreme is the absolutely *abstract*, and an ideal which, properly speaking, has no existence, but rather only possibility.[4] Closer to home is the *probable*, such as the likelihood of rain in the weather forecast. The probable is an 'actual possibility'.

VIRTUAL THINKERS: PROUST, BERGSON, DELEUZE

As the quote from Proust suggests, there is a history of reflection on the virtual. Three key authors stand out: Proust, to whom the definition is attributed by Henri Bergson, the second figure, and Gilles Deleuze, who attempts to recast Bergson's intuitionist style of thought as a general approach (for intuition is not properly a faculty but a methodology).

Although his work is much referenced in philosophical discussions of virtuality (cf. methodologically: Deleuze, 1988; Badiou, 2000), Bergson rarely uses the term *virtual* himself. Gillian Rose argued that Deleuze offers not *Bergsonism* but a 'new Bergsonism' (Rose, 1984: ch. 6). It is more common among English commentators on Bergson to make no mention at all of the virtual until the late 1980s (cf. Pilkington (1976), notably the discussion of Bergson and Proust; Kolakowski, 1985). We cannot therefore go back to an authoritative definition or philosophical discussion of 'the virtual', but are left to our own devices. For example, in *Matter and Memory* (Bergson, 1988), the virtual is used only as a descriptive term, an adjective which helps summarize a much longer (and now outdated in terms of both the language of realization (see below) and in terms of neurophysiology) discussion of stimulation, perception and memory. There are important literatures and long-running debates on the writings of all three. This section merely glosses some of the key points of the virtual in relation to each and indicates key interpretations and texts. Philosophical positions are made more complex by the lack of a sense of intellectual development over time (their positions change and develop) among adherents of each figure and the tendency to isolate the virtual as a philosophical issue rather than locating it as a key problem in everyday life and affairs.

> in other words, the virtual image evolves toward the virtual sensation and the virtual sensation toward real movement: this movement, in realizing itself, realizes both the sensation of

> which it might have been the natural continuation and the image.
>
> (Bergson, 1988: 131)

Bergson argues that the (human) mind establishes a gap between stimulus and response which enables remembrance of experience (memories similar to virtual images in optics), if in a rather passive manner, and thereby opens the possibility of unpredictability and freedom. It is only the mind that integrates the multiplicity of our worlds into a unified flow of duration (*durée*). This unity is indiscernible in analytic methods based on the division (or 'spatialization') of time into moments (Deleuze, 1986, II: 81–82). This allows Bergson to suggest that actualization can be turned back on itself, a philosophical move which is criticized as a replay of the metaphysical quest for a unity of all things (see Douglass, 1992; Badiou, 2000). Objects are 'the point of indiscernibility of two distinct images, the actual and the virtual' (Deleuze, 1986, II: 82; Deleuze, 1994: 209–210). A full examination requires the analysis of both sides of an object or situation – a kind of 'double circuit' which I would expand to a fourfold optic which considers all four ontological modes.

> If everything about matter is real, if it has no virtuality, the proper 'medium' or milieu of matter is spatial. While it exists in duration, while clearly it is subject to change, the object does not reveal itself over time. There is no more in it 'than what it presents to us at any moment.' By contrast, what duration, memory, and consciousness bring to the world is the possibility of unfolding, hesitation, uncertainty. Not everything is presented in simultaneity. This is what life (duration, memory, consciousness) brings to the world.
>
> (Grosz, 1999: 25)

The virtue of conscious subjects is that they reverse the virtual to actual sequence of becoming. 'A body becomes virtual by organizing itself into a subject . . . this virtual effect then posits itself as the actual ground' (Colebrook, 1999). Bergson's dualistic version relates the virtual to the actual (rather than the concrete) establishing a series of binaries:

Actual – Virtual
Matter (Object) – Memory (Subject)
Present – Duration (Progression)
Spatial (Synchronic) – Temporal (Diachronic)
Non-Organic – Living
Inert – Potential
Complete – In-process

The relation of influence between Bergson and Proust is much debated (for a refutation see Pilkington, 1976: ch. 4). Therefore, we cannot simply assume that all three share the same definition of the virtual, even though Bergson and Deleuze repeat Proust's mantra. Further problems arise in translation. For example, in the translation from the French '*actuel*' used by Proust and Bergson, the notion of the present is lost while the actuality of the probable is muddied. For this reason, and in accordance with more recent scholarship such as Stivale and Hardt's (above), I argue that the terms need to be further clarified by resetting the dualisms favoured by Bergson in particular, ideal–actual, existing and non-existing in a mutually exclusive manner. It helps to view these terms in tabular form, a tetrology of the real and possible (Table 2.1).

- The *virtual* is a 'real idealization' such as a memory, dream or an intention.
- The *concrete* is an 'actual real' such as a taken-for-granted thing, an actualized idea and anything that embodies memories. It is the event, our everyday 'now'.

- The *abstract* is a 'possible ideal' (expressed as pure abstraction, concepts);
- The *probable* is an 'actual possibility' usually expressed mathematically, such as a percentage.

Table 2.1 The virtual and the concrete

	Real (existing)	*Possible (not existing)*
Ideal	virtual (ideally real)	abstract (possible ideal)
Actual	concrete present (actually real)	probable (actual possibility)

The best contrast to the virtual is the concretely present (which may also be called the real actual).[5] *The virtual is distinct not only from the concrete, but also from the abstract.* This is a continuum of soft oppositions; for example, the virtual might feed and nurture the possible and is clearly in a dependent relation to the actual (in the case of virtual reality, this would be exemplified by its reliance on telecommunications infrastructure, technology and living bodies). Is the past real? Yes, virtually, inasmuch as there is an actual past of events which were once the concrete present and which are now really existing memories, cognitive representations reconstructed each time we remember (Neisser, 1982; Antze and Lambek, 1996; and see below).

Where it does appear in the work of Bergson, the virtual is entangled with duration as part of his study of the importance of the subjective understanding of the flow of time (*durée*). This true time is grasped only in the course of its actualization, a process of differentiation and creative evolution rather than the production of concrete instantiations which were already established virtually. That is, the concrete is not a copy of the virtual – the relation is not one of resemblance or identity (Deleuze, 1994: 212; see below) as it might be between the concrete and abstract representations such as concepts or images. Bergson designates this process as a vital force (*élan vital*).

Difference is primary, a factor which Deleuze exploits a generation later to develop a post-structuralist philosophy of difference. Thus, across different categories (for example, race) Deleuze provides a conceptual toolkit for understanding their commonality-in-difference, with the proviso that he is not providing a new form of metaphysical unity.[6] Others, such as Butler, attempt to capture something of this relationship by developing the idea of 'performativity' as creative 'citation' (Butler, 1993).

> Different levels only coexist insofar as they remain virtual (at the level of essences). What coexisted in the virtual ceases to coexist in the actual . . . cannot be summed up . . . each one retaining the whole, except from a certain perspective, from a certain point of view. These lines of differentiation are therefore truly creative: They only actualize by inventing . . .
>
> . . . The whole is never 'given' . . . it cannot assemble its actual parts . . . an irreducible pluralism reigns.
>
> (Deleuze, 1988: 101, 104)

Deleuze argues that the virtual is constitutive but ineffable. It is not opposed to the real and is therefore not realizable in the same way that the (non-existing) possible is. Axiomatically, the possible is an image of the real, a negation. Realization is a process of bringing the possible (the abstract or the probable) into existence in a manner that resembles it. In contrast, the virtual is fully real but can be actualized as the concrete. For Deleuze, 'the actualisation of the virtual . . . always takes place by difference, divergence or differenciation. Actualisation breaks with resemblance as a process no less than it does with identity as a principle. Actual terms [the concrete] never resemble the [virtual] singularities they incarnate. In this sense, actualisation . . . is always a genuine creation' (Deleuze, 1981: 125).

In his influential *Becoming Virtual* (1998), Pierre Lévy turns from history, memory and the past to apply Deleuze's work with a focus on imaginative and artistic creativity in the contemporary moment. His interest is in the relationship between the becoming of new creations and ideas, which Bergson argues can be apprehended only subjectively, and the event in which the 'new' takes concrete form (Lévy, 1998: 172). These two axes are also axes of time and space, with the virtual flowing towards actualization on the former (time), and the possible which takes on substance to become real on the latter (space). These axes are not only non-exclusive but parallel and complementary. They both describe the same dualism which lurks in the background: existing–not existing (Lévy, 1998: 171).

Drawing on the work of Deleuze's writing partner Felix Guattari (Guattari, 1992), Lévy also considers the process and risks of virtualization, a process of creative enquiry and questioning which opens up problem frames to critically question cultural formations, 'the way things happen'. Virtualization moves from situations (*'l'actuel'*) to create problems 'the knot of constraint and finality that inspires our acts. Final causes, the "why" of the situation' (Lévy, 1998: 174). Again, however, Guattari sets up his discussion in terms of a matrix of two dualisms: the Real and the Possible, and the opposition of the Virtual to the Actual (*l'Actuel*). The result is to exclude from the internal categories of the tetrology above a category of the concrete and material.

Although Table 2.1 is enormously simplified and provides a non-exhaustive list, the analysis is useful for teasing out the characteristics of often taken-for-granted concepts, which are key to our understandings of culture, and of those fears and hopes, which underlie our outlooks and motivate our actions. Of course any ongoing action, belief system or argument mobilizes all four facets of ontology. No one would conclusively win an argument on the basis of complete abstractions; there will be attempts to test even the abstractions of theoretical physics against reality. This is not because every truth is empirically testable[7] but because there is both a material benefit to extending theory towards controlling the real, and something akin to a compulsion to extend our understanding across the full range of ontological facets. Hence one marshals evidence (the concrete), chance and coincidence (co-variance and probability), and abstract ideals (moral values) in the assertion of regularity and of laws of social action (virtuals). Thought takes us beyond the present moment of the actual, not only to abstract ideas but to general problematics, to the historical and to the realm of principle, all of which are virtual.

In social science, for example sociology, attempts are made to interpret actual events. The stories constructed by sociologists and anthropologists are argued to be more than mere abstractions, and more than statistical predictions – they are held to convey something that is really taking place 'beneath' the surface of events.[8] Thus, for example, an uprising may be interpreted and argued to be a manifestation of class tensions around economic entitlements. Understanding global economic relations as intangible but powerful 'virtual' relations frames our attitudes and actions towards national economic instability and the popular experience of change in the job market. 'The virtual' becomes a template for understanding and reacting to events in everyday life whenever societies face a situation in which distant events (a corporate merger) have local impacts on a related but quite a different register (prices for a service).

The real qualities of the virtual, such as a memory of an event, distinguish the virtual from the unreal, or even surreal, qualities of the abstract. But the strength of Table 2.1 is that it allows us to both distinguish the virtual from – and relate it to – worlds of material existence, the mathematical worlds of probability and possible occurrences, and the abstract world of pure idealizations. These relationships are mediated by human agency, the flow of time and concurrence of place – something that is captured in the everyday language of surprise at transformations, the calculation of risk and the invoking of spirits. A risk or myth, an event or dream draws on all aspects of the real and possible. Contemporary cognitive science and neurology shows Proust to be incomplete: in any dream one could find not only the virtual but the concrete present of neurochemistry, hormones and the electrical exchanges of brain cells. A caution against reducing to one element or another is therefore in order. None the less, the table has an analytical and heuristic value: we can learn by considering social action in terms of each of the four aspects of the tetrology and in terms of their exchanges with each other. Walter Henry in a trenchant analysis points out that all communication involves the concrete (voice, inked letters), the virtual (coded meaning), the abstract (ideas), and the probable (author's intention) (Henry, 2001). These categories are woven together in everyday cognition and interaction. Thus it is not a matter of drawing on one single category – we rarely find pure examples of the virtual – but an assemblage of the terms. This explains how in imprecise everyday speech it is often difficult to demarcate where a naming of materiality, such as a useful product, stops and a projection of probability, shaded with abstract belief and glossed over with virtualities such as a brand name begins.

As the case of the Eucharist suggests (above), attempts to invoke the spiritual, for example, involve moving the virtual into the concrete, and giving abstract ideas the force of a material presence. The unpredictable is defined and the invisible may be divined in such rituals. We greet these shifts between the categories of existence with surprise and awe, understand them as miraculous events. Examples of such movements between categories may be added to flesh out the tetrology (Table 2.2). Deleuze speaks of actualization as a dramatization that enacts a simulation rather than a copy of an original image (as in the case of realization – Deleuze, 1981: 216–220). It is a 'contraction' of virtualities, which comes into being through an indexical leap rather than continuity with an original. Yet the virtual continues to inhere 'within this actual dispersion as that which both constitutes it and into which it dissolves' (Widder, 2000: 129; cf. Deleuze, 1993: ch. 7). There is thus an axis of realization between the possible and real, and an axis of actualization between the ideal and actual that are characterized by very

Table 2.2 Figures of speech and movement between categories of the real and possible

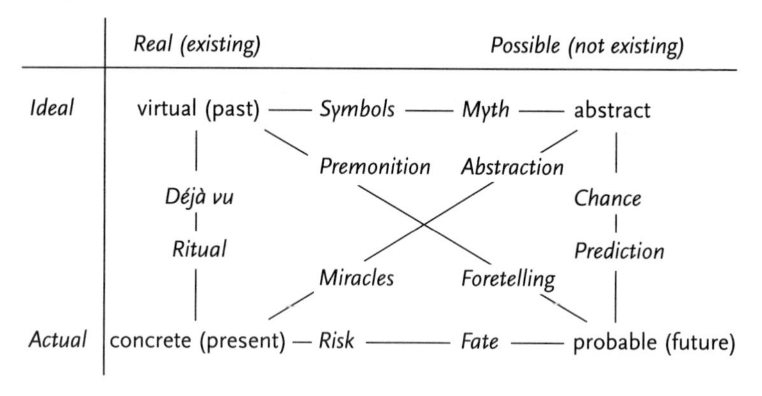

Virtual↔Concrete

- *Déjà vu* exemplifies the sensation that the present has already been experienced in a dream. The actual–real present is lived in a surreal, dreamlike state as virtual, or ideal–real.
- *Ritual* actualizes latent possibilities, conjures the past with a view to altering the present.

Virtual↔Abstract

- *Symbols* represent and thus make present abstractions by giving them a form.
- *Myth* formulates the past as an idealization, purifying it of factuality in favour of moral and ethical ends.

Abstract↔Probable

- *Chance* is the abstract idea of the play of probability.
- *Predictions* formulate the abstract ideals into calculations of the actually possible.

Concrete↔Probable

- *Risk* is our pragmatic approach to probability – we take risks on the chance that a computable probability will not actually occur.
- *Fate* describes a present or an outcome as a future prescribed as an actuality.

Concrete↔Abstract

- *Miracles* are said to occur when the non-existing ideals suddenly materialize.
- *Abstraction* conceptualizes the concrete present as a pure (non-existing) idealization.

Virtual↔Probable

- *Premonitions* are visions of probabilities in the felt form of emotional sensations. They are real idealizations of actual possibilities.
- *Foretelling* the future casts a calculated, possible outcome as something that has already been conceived, imagined and possibly represented.

---

different relations. Above all the performative relations of actualization challenge us to rethink identity relations characteristic of the process of realization

Similarly, there is a range of traditional 'beings' who 'figure' these exchanges. These figures are our contemporary pantheon of spirits. Perhaps we most fear the figures of the past: *ghosts* and apparitions are virtual (they are reputed to exist but have no material substance). They cross into actuality as manifestations that have actual effects (hauntings). Biblically, *angels* appear in dreams to foretell the future (probable–virtual). Angels as messengers, witnesses and guardians provoke chains of actual events, connecting the divine and concrete. To this day, Europeans perceive outlines of Greek mythical figures abstracted in the constellations (concrete–abstract). The figures of fate and time preside over secular, modernist mythologies of progress. And the almond-eyed silhouette of the *alien* has moved what were once called 'creatures from outer space' from fantastic abstraction into being a figure of the probable, a being whose existence is debated as much in the virtual mode of recovered memory as in the fictional television series *The X-Files*.

Flows between categories are not only a matter of analysis. In many cases, *rituals* are used to shift group perceptions and understandings in a coordinated and actual manner. While we have argued that the virtual has always been a crucial part of ontological thinking, even if it has been less recognized in some periods and cultures and more in others, what historical examples of the virtual can be found outside of historical records of intellectual thought and scholarship? The historical importance of the virtual may be detected from records of ritual events and ceremonies; for example, the coronation of kings and queens bestows a title and ascribes an identity to an actual individual. Historically, royalty were understood to be god-like beings – the Japanese Emperor continues to be honoured in these terms. Coronations *actualize* the virtual, bringing the idea of 'the King', for example, down to Earth in the form of an actual individual. The transformation from, for example, 'Crown Prince' to 'King' is engineered via an elaborate ritual in which social attitudes and expectations are shifted and bodies move ritually from one status to another.

In day-to-day terms, distant loved ones or those who travelled and emigrated to far-off destinations in the past were removed from the face-to-face interactions of family and tied to home only through the occasional delivery of precious letters by post. While clearly still existing, distant relations are hardly a material factor in everyday life. Rather they are virtual and abstract, for distance and lack of contact remove them from many day-to-day considerations. In a deep sense, the face-to-face world of communities and clans was never totally governed only by the real–actual because humans, family groups and communities have likely always travelled and been nomadic to a certain extent. Those who were absent were much like the constellations – more abstract than virtual – and all the more so because they did not figure in the highly localized routines of everyday life during their absence. *Faith* kept not only memories alive of those travelling away, sailing perhaps great distances, but describes a key mental activity in any form of the virtual: the willingness to accept an ideal 'in essence' in the place of actual presence, seconding the virtual as actual.

From the eighteenth century until the early twentieth century positivism was the unchallenged theory of science. It insisted on the observability of all that is 'real' dismissed the virtual along with the abstract. Such nominalism has now been shown not to have been, nor to be the case in scientific practice (cf. Latour, 1993). A rigid division between the materialism of empirical reality and the idealism of abstract thought simplified the ontological into two categories: existing versus non-existing. The virtual was conflated with abstraction. Probability was largely misunderstood, kept out of the picture by a focus on only the present and on only Euclidean space. The virtual was thus suppressed, although, historically, many cultures have long regarded the virtual (memories, dreams and so on) as a significant form of the real. Religious and mystical rituals were often intended to actualize the virtual – injuring a voodoo doll would cause physical harm at a distance.

In earlier times, the virtual was considered an important aspect of the means of awareness and knowledge. Liturgy and ritual consciously evoked the virtual as real, not trivial. Spiritual realities and truths are approached through the virtual and the abstract. Religions still provide countless miracle stories of the abstract becoming concrete. Visual images of the abstract operate as virtual forms in which the possible but not actual may be understood as 'an ostensibly sensory phenomenon' (Nie, 1998: 116). Rather than a descriptive or propositional knowledge of the world, the spiritual is offered as a transformative form of knowledge which is expressed imagistically as pre-verbal figures laden with affect, paradox, the coalescence of opposites, and embodied, sensory forms of knowledge:

> As literary constructs, the miracle stories use imagistic dynamics in material objects to stand for and evoke a real – if imaginal – experience in the reader or listener of a leap mimicking that of the . . . 'divine'. In the ritual situation, this enactive experience of the leap would be the basis of a belief that would store this image as a model for subsequent [religious] perception. In

> addition, it could establish the predisposition to foreground the meditative-affective – that is, imagistic or 'dream' – mode of awareness . . . in situations within the religious context.
>
> (Nie, 1998: 116)

With the rise of computer-mediated communication and digitally created virtual environments, the virtual returns to 'Western' cultures (see following chapters) in the form of absences made present. Simulations offer virtual environments – clearly they are 'something' but there is no materiality there. This has puzzled theorists used to thinking in the either–or terms of the material–idealist division which, like a flute with only one note, offers only notes and rests: the materially existing or the abstract, non-existing. Although its name may be unfamiliar, the virtual will be shown to be a category with which people are comfortable and in terms of which they live their lives. In part, this is because the virtual has always existed in its 'traditional', ritualized forms.

Recognizing the virtual allows us to see more clearly how it enters into categories of thought and into descriptions of the world and embellished narratives which cast the world in a given light or frame problems. Moving away from the simple actual–ideal binary of (concrete–abstract) to recognize how the virtual fits in relation to these terms and with probability has important implications for the way in which risk is understood and managed at a societal level.

MEMORY

The virtual includes those elements, such as memories, which are not simply abstractions but are *real ideations* (day-dreams, the past and so on). They may be experienced as real, but they are neither tangible nor actual. This is not merely a matter of semantics but essential to understanding the significance of simulations and their relationship to material reality. The actual is context bound. Memories are by definition virtual. They have to be worked up each time they are recalled or invoked (Neisser, 1982). The virtual is always real, even if it is a memory or a past event, but it is not actualized in the present except via specific human interventions, such as rituals, which make these memories or other 'virtualities' tangible, concrete. The psychological, museum and archaeological metaphors by which the past is conceived

> tend to transform the temporal into the spatial and are intensely visual. Layers are excavated, veils lifted, screens removed. As such the recall of socially and effectively charged events involve a social organisation of a present space (structured encounters with a site, even tours or processions), with specified stopping-places and actions (ablutions, obsequities, gestures and readings) and time (seasonal ceremonies), as well as an historical space and time (Kirmayer 1989). Memory is reconstructed anew each time through secular rituals of for example the systematic, often guided, tour in which the site is 'framed' by discourse. The position of the viewer may be left in question or explicitly positioned, but there is always a space, a distance, between the spectator and her memory.
>
> (Antze and Lambek, 1996: xii)

This distance is the space of '*metaxis*': the operation of the imagination which connects the perceptual environment with the virtual and abstract world of meanings which over-code our perceptions. Psychologists have advanced the suggestion that 'there is a survival advantage to constructing a social reality that corresponds to some objective reality' (Shapiro and Lang, 1991: 689). However, this naturalizes an either–or dichotomy between the 'real' and the 'fictional' which even psychological research indicates is too crude. A review of the literature confirms some of our categories: Dorr found that as children mature into adolescence, their definition of 'real' shifts in a complex manner from 'something fabricated but "possible"' to something fabricated but probable or representative (for a review see Dorr,

1983). Thus children may believe in fairies (abstract, that is, improbable but possible to imagine) while adolescents (and many adults) deny their existence but idolize the screen characters portrayed by actors. These dramatis personae are probable and virtual but not actually existing characters based on the recognition that the acting is lifelike. Thus we may learn selectively from fictional scenarios (novels) as much as from life experiences, making quite sophisticated judgements about what is relevant to import into our everyday actions (Tyler, 1984; Rubin, 1988) – or falling victim to urban myths and media hoaxes (such as the 1938 CBS broadcast of *War of the Worlds* (see Fedler, 1989)) depending on variables such as the level of anxiety, urgency of the situation in which we find ourselves, trust in the source, and pressures to conform to socially accepted beliefs and attitudes.

Earlier, it was noted that children's sense of reality includes the possible as well as the real, narrowing down to exclude abstract entities such as mythical fairies and distinguishing fictional events and tales from historical happenings (Dorr, 1983). None the less, the past never recurs literally; it has a virtual existence as a narrative, a memory, an ideation (see above). How does the social commemoration of these virtualities interact with children's simultaneous quest for certainty and the sense of control that authorship of their own imaginative fantasies brings? What is the impact of digital virtuality, which offers a magical, computer-mediated version of the global village?

Fiction and fantasy reveal an enduring adult willingness to believe in virtual and abstract entities, especially where a story of reasonable doubt in the empirical facts can be constructed. Ambiguity is fascinating. How many shooters were responsible for the death of US President John F. Kennedy? A host of films and television specials have been made on this topic. The Ottawa Elvis Sighting Society may believe Elvis is alive – even though most would admit that it is unlikely he would ever have moved to the capital of Canada, a place that meteorologists have shown

is actually, materially, the coldest capital city in the w
preference for materialism emphasizes the actual–real,
virtual is a required category for distinguishing non-existing and
ideal abstractions such as concepts from ideal but really existing virtualities such as memories and myths (see also Antze and Lambek, 1996) and, of increasing importance, simulations and totalities such as groups and classes. These exist in and of themselves, but are not actualities, again, not concrete.

TECHNOLOGIES OF THE VIRTUAL

Howard Caygill comments:

> the ensemble of techniques that make up the world wide web and its technological basis in the interlinked servers of the Internet seem to promise a new art of memory in which knowledge as technological invention replaces knowledge as recollection, and in which the archive appears as an effect of the links made possible by the technological work of memory rather than a given (and carefully policed) store of information.
>
> (Caygill, 1999: 2)

In effect memory moves from the virtual realm of recollected knowledge into the material realm of stored and access information.

Techniques of the virtual create the illusion of presence through props, simulations, partial presences (such as a voice conveyed by telephone or thoughts written in a book) and rituals which invoke the past and make absent others present. They aid metaxis from the virtual to the actual by giving concrete presence to intangible ideas. Historically, a growing web of communications, beginning with the early couriers and envoys, first between the courts of rival countries and empires, later across those empires in the form of postal systems which served common people, and finally via telegraph and other forms of telecommunication to the far corners of the globe, culminating in the spread of the Internet and email as a grass-roots alternative to phone, fax and telex which girdles the globe (even if this is more the exception than the rule in some developing countries – see Chapter 4). In the twentieth century, 'the decoupling of space from place accomplished through the use of the telephone implies that "virtual" life has been coming for a long time' (Hakken, 1999: 90) – and has been with us even longer if the brief historical survey we have given is correct.

Perspective, used in images since the Renaissance, is one such technology. It is a convention for representing scenes, and giving representations the appearance of being virtually real. Rather than being strictly accurate, perspective creates the illusion of space of a two-dimensional surface for fantasy and contemplation. Effective representation of a hand outstretched towards the viewer requires alteration of the rules by 'foreshortening' the length of the limb. The secret of great classical art is often how it breaks the mechanical rules of perspective: for example, Michelangelo's sculpture of the *Pieta* (there are actually several versions and studies). In this marble sculpture of the dead body of Christ held by his seated mother Mary, Michelangelo reduced the size of Christ to fit him across Mary's lap, in the name of obtaining an integrated composition. Whether Christ is tiny or Mary is huge – the point is that either way, the sculpture is not simply an image of actual bodies (as in an abstract representation); it is a virtuality.

The realist preoccupation with simulating the material world is defended on geometric grounds, but static perspective compositions composed from a single point are not the 'natural' way of seeing things. We move around and scan the environment with both eyes, binocularly (much discussed since the 1960s – see Hillis, 1999). Even if it is so culturally entrenched that it seems natural, perspectival technique has developed over the centuries to provide a rationalized image for the viewer. Perspective is an odd form of special effect aimed at arresting vision.

Perspective becomes more than a technique; 'seconded' perspective becomes naturalized as a way of seeing that dictates a visual approach to the world in which we are preoccupied with space and geometric alignment. Time, lost without the sequentiality of movement, is rendered virtual with the effect that images come to be frozen snapshots of some point always in the past. 'We are persuaded by this theory, to view the world from a single fixed position, with a single lead eye', even though we break the rule in the virtual: 'In dreams, objects and people often appear to be "there" and "here" at the same time. Roads can lead to more than two places at once. Spaces can be logically unrelated but appear connected nonetheless' (Wachtel, 1980: 84–85).

SUMMARY

This chapter elaborates on the definition of the virtual as '*that which is so in essence*' but not actually so. This notion of the virtual as essentiality was contrasted with the common distinction between the virtual and the real. It was argued that a better contrast opposes the virtual with the concrete. A four-part definition of the virtual, the concrete, the abstract and the probable was proposed. The virtual is ideal but not abstract, real but not actual. It is ideally real, like a memory. Of more significance is the weaving together of these ontological categories in our representations of reality, of the past and of the future. Virtual elements are embedded in everyday activities and the language we use. Ritual, miracles, understandings of risk and fate all involve slippage between the categories as the virtual is actualized, the probable takes place – as our fears and dreams 'come true'.

We examined the roots of the virtual in the everyday mental ability to accept the 'almost so' in place of the actually so. *Metaxis*, or the ability to imaginatively close up the gap between fiction and reality and between the virtual and the actual has a long history, and may be found in social rituals and in the historical record. Historically, cultures have long regarded the virtual (memories, dreams and so on) as a significant form of the real. Rituals were developed to invoke and manage virtualities, integrating them into life as carnivals, sacred times and places, and mysteries. In some cultures the virtual was easily mistaken as concrete. However, the virtual was challenged by modern positivism, which dismissed it as non-existing abstraction and concentrated on the concrete and the probable.

Computer-mediated communication reintroduces virtualities as important presences in the form of distant but significant others – friends, clients, teammates – and in the form of digital simulations for play and by which future trends and actualities are anticipated and prepared for. Concrete techniques of the virtual including not only computers but also conventions such as perspective, support the virtual and give it tangible presence. The embeddedness of the virtual appears in approaches to problem solving including the frames within which we pose questions or understand problems. Slippage between the virtual and actual appears to have been widely accepted – we embrace virtual substitutes while nostalgically remembering (i.e. virtualizing) what we might call 'the real thing'. The following chapters will consider the rise of digital virtualities in the form of online virtual environments; the role of virtuality in videogaming and simulation; the impact of digital virtuality at work; in the economy; and in personal lives with special attention to risk and security.

3

DIGITAL VIRTUALITIES

This chapter considers the rise of simulation software and hardware as a digital form of the virtual. From the painted circular panoramas of the 1800s to immersive virtual reality and digital renderings of environments in role-playing and other online games, there is a long history of virtual environments. Central to the recent history is the rise of sophisticated graphic display hardware and software, complex geographical information systems (GIS) and the popularity of video- and computer-games. This chapter outlines:

- The history of virtual environments, dating from the panoramas of the 1800s needs to be separated out from virtual technologies and simulation technologies.
- The 'liminoid' quality of virtual environments, 'betwixt and between' people.
- Computers as filters at the boundary of the concrete and a digitally created virtuality.
- The autonomy of cyberspace.
- Impacts on everyday life.

Time and again a number of late twentieth-century films and novels popularized the idea of digital virtual reality, known by its popular acronym VR. Of these, the most unusual film, *The Matrix*, presented everyday life as the simulation. In the film everyday life is a virtual reality; a hoax perpetrated on a sedated population whose bodies are banked by a robot race to harvest the humans' meagre electrical energy. Coyne comments:

> to [the] . . . catalogue of testimonies to the ineffability of the real we can add the concept of cyberspace. There is no technology that enables brain implants linking us to a data matrix, and as phenomenology tells us, experiencing and knowing do not function that way anyway. Cyberspace is an imaginative fiction that provides a stand-in, a substitute, or a wild card, in various digital narratives.
>
> (Coyne, 1999: 225)

In the context of digital technologies and their social forms at work and in the telecommunications of advanced capitalist societies, 'virtual' comes to equal 'simulated'. Rather than being something which is an incomplete form of reality – something real in essence, 'almost', 'as if' and offered as being 'as good as' – the virtual comes into its own as an alternative to the real. The virtual is not merely an incomplete imitation of the real but another register or manifestation of the real. In some cases it is better than the real. Although often slowed by inadequate transmission capacity, the high-speed networking of computers through the Internet can allow the seamless exchange of information such that matching interactive environments can be created at either end of a computer-mediated communication. Virtual environments involve the construction of a simulated shared space 'in the wires', so to speak. For each participant, the other participant(s) and their gestures are displayed within a computer-drawn shared environment such as a room, which can be explored by shifting one's point of view.

During the 1980s and 1990s, simulation software for

representing three-dimensional objects brought the virtual to prominence in digital form. These digitally generated environments are virtual in part because they have no location in the actual world, but rather depend on the ability of users to imagine virtuality (see Chapter 1) and the artistry of graphic software and computer interfaces (Rothenberg, 1993). Virtual reality duplicates 'reality' by means of technology. It provides a simulation in which to experiment with substitutes of the material world that are 'close enough to the real that its conditions may be tested without the normal risks. In these cases technology provides prostheses for the real in order to better control it' (Poster, 2001: 129).

In simple, widely available systems of the late 1990s, such as Apple's QuickTime VR (virtual reality), one could turn in different directions and sometimes track forward to a new position using the arrow keys on a computer keyboard. Head-mounted displays and datagloves could sense the user's movement and pan or shift the display accordingly to give an illusion of immersion within the computer-generated space. This went a step beyond video-conferencing where two cameras are linked by phone wire or Internet and broadcast to each other, putting the 'others' whom one is conferencing with firmly 'in' the television monitor and yet leaving each participant outside in their own local milieu.

SIMULATION

Simulation was already an issue in the case of studio-produced music in the 1950s. By the 1970s, many of the musicians would have heard only the track they recorded, until the point when all the tracks were mixed together. Writing in the benchmark (1990) text, *The Mode of Information*, Mark Poster gave electronic music reproduction as an example of simulations where 'no longer is there an original performance, only separate performances of tracks: the performance that the consumer hears when the recording is played is not a copy of an original but is a simulacrum, a copy that has no original . . . rock performances exist *only in their reproduction*'. Rather than a direct translation or representation of a sovereign reality, 'The electronic mediation' of musical information dispenses with 'the original performance' in such a paradoxical manner that he worried, in line with poststructuralist theory, that it subverted

> the autonomous, rational subject for whom language is a direct translation of reality, instantiating instead an infinite play of mirror reflections, an abyss of indeterminate exchanges between subject and object in which the real and the fictional, the outside and the inside, the true and the false oscillate in an ambiguous shimmer of codes, languages, communications. In this world, the subject has no anchor, no fixed place, no point of perspective, no discreet centre, no clear boundary. . . . In electronically mediated communications, subjects now float, suspended between points of objectivity, being constituted and reconstituted in different configurations in relation to the discursive arrangement of the occasion.
>
> (Poster, 1990: 11)

If this was the worry over electronically reproduced music, imagine the panic that ensued in the case of digital media!

Acoustic telephone space

In the Introduction we considered the virtual in examples from the 1500s to the 1800s. The telephone has long intervened in our sense of the world as a space of distance by providing virtual auditory spaces in which, generally, a person in one place is brought into earshot, so to speak, of a person in a distant place (see Ronell, 1989). Although not necessarily digital, telephone conversations can convey a sense of presence and intimacy with another person far away – a sense which can only be called virtual.

Calling a telephone conversation a type of virtual space forces us to re-examine the oddness of the idea of a virtual space imagined to be enduring and independent of geographical spaces. It also takes the concept of the virtual away from the overwhelmingly temporal emphasis given to it by writers and thinkers such as Proust, Bergson and Deleuze (see Chapter 2). If there is any single cleavage in discussions of the virtual it is between the continental notion of virtuality being actualized in time, and the Anglophone and Asian notion of the virtual space or environment. The spatialization of communication as a multi-variable environment rather than a bipolar line of exchange back and forth between two callers comes with addition of the visual. While futurists long anticipated 'videophones' this was rarely conceived of as a full-fledged environment, merely an animated image to go with the speaking voice. 'The virtual' is imagined as a 'space' between participants, a computer-generated common ground which is neither actual in its location or coordinates, nor is it merely a conceptual abstraction, for it may be experienced 'as if' lived for given purposes. As Bogard points out, virtual spaces cannot properly be said to be in the same locale as one or other of the participants. Virtual spaces are indexical, in Pierce's sense, in that they are interstitial moments (see also Elmer, 1998). The virtual involves a modification of understandings of locatedness and the relations between distinct places and of inside–outside relationships, and specifically of the disappearance of the outside, and of outsidedness, as part of new spatializations and iconographies of social interaction (see Deleuze, 1988: 74–79; Colombat, 1999: 203; Rodowick, 1999: 39).

The virtual is liminal, 'betwixt and between', a threshold (*limen*) between at least one immediate lived milieu and the distant ground of the other(s). In it, everything is representational, a convenient fiction by which participants 'meet' but only figuratively; elements interact 'in essence' but not physically (see Chapter 1). Beyond the transmission, bricolage and the animation which is the labour of the technologies involved, there is always an innately human work of *metaxis*, translation and imagination which transposes digital action and virtual encounters to the world of living animals and objects.

This spatialization extends beyond understanding that digital domains will be treated as virtual spaces;[1] it includes cooperation in the treatment of these spaces as serious domains of action with an equivalence status to face-to-face, embodied interaction. Part of the necessary performative competence is an acceptance of the conventions mapping the virtual and the concrete on to each other and organizing the labour of supporting this meshing of digital virtuality with the embodied interactions and logistics of the world of bodies and concrete things. This is embedded in organizational procedures, such as accounting, and enacted through a complex of institutions by which not only communication and computing technology are extended by bodies marshalled in organizational spaces (such as a bank branch or a store) but also by specific interfaces such as a computer terminal or a cashier's workstation (see Chapter 6).

Virtual spaces have an elusive quality which comes from their status as being both no-place and yet present via the technologies which enable them. However, just as these environments are not spatial per se, but only virtually so, they also have duration but, strictly speaking, neither a history nor a future. Of course there is a history of virtual spaces and of the technologies that make possible the transposition of interaction away from the limits of the human body into various media. Inside a virtual space itself there is only the immediacy of the scenario displayed. This 'presentism' (Maffesoli, 1996) temporalizes virtual space, making it, and processes or events in it, something that always happens 'now', in the present. Although they can be archived, creating a form of virtual history, both virtual space and virtual objects are merely retrieved and re-created in whatever present moment one may choose to witness them. One may go 'back' to a previous web page or virtual 'room' but one may also 'jump' as far back or forward as one wishes. A sense of elapsed time must be accomplished by developing a spatial narrative of the path that one has taken and which may be retraced. Researchers in the United Kingdom's 'Virtual Society?' Research Programme have argued that 'the ICT industry works with axiomatic ideas about memory as storage (of data and of the means to access data)'. However, 'what counts as adequate remembering?' (Harvey *et al.*, 2000) is a question answered in advance by a rationale geared to the predefined needs of software functionality, not remembrance or *reverie*.

Perhaps there is a 'gut' recognition of this distinction. While software has been created to provide timelines and 'virtual tours' of historical sites (McCarthy, 2000), there is no virtual Auschwitz. Virtual memorials focus on testimonial and eulogy text over monumentality (see e.g. the Virtual Vietnam Memorial at *http://www.VirtualWall.org*). These often include testimonials to a person or a form of 'Visitor's Guest Book' commentary on the power of physical monuments, or of *remains*, which the virtual site supplements but does not supplant.

Cyberspace novels

The ambitions of cyberspace are much grander, however. 'Cyberspace', coined and popularized in the science fiction writings of William Gibson, is a global simulated environment accessible by an almost 'transparent' neural interface. Cyberspace is 'a consensual hallucination' as Gibson called it (Gibson, 1984: 67). This virtual environment is imagined as a virtual world, an alternate computer-generated reality of telecommunication networks of economic exchanges and databanks holding all economic exchanges. Stone argues that his highly influential novel, *Neuromancer*,

> reached the technologically literate and socially disaffected who were searching for social forms that could transform the fragmented anomie that characterized life in Silicon Valley and all electronic industrial ghettos. In a single stroke, Gibson's powerful vision provided for them the imaginal public sphere and refigured discursive community that established the grounding for a new kind of social interaction . . . *Neuromancer* . . . is a massive intertextual presence not only in other literature

> production of the 1980s, but in technical publications, conference topics, hardware design, and scientific and technological discourses in the large.
>
> (Stone, 1992: 95)

The notion of cyberspace was always more than simply an environment. It describes the type(s) of social world that VR might afford. Although it remains largely a fictional construction, the term has had a huge impact on the popular imagination – as the technology has developed, science fiction enthusiasts have been waiting with very imaginative ideas for the uses of VR technologies (Valente, 1995: 314). Gibson's cyberspace is depicted as a world of unequal access to data, controlled by corporations and sought by trespassing hackers, criminal 'data cowboys' and others engaged in industrial espionage. While it is presented as an addictive sensory utopia, its contents are just the reverse, a dystopia of privatized, alienated data defended by passwords and debilitating electric shock fields.

The significance of Gibson's vision of 'cyberspace' was that it provided an organizing image and cognitive mapping in which researchers, entrepreneurs, hackers and weekend Netsurfers could recognize themselves as a community on the model of an imaginary city (Fitting, 1991: 311; Stone, 1992: 99). For example, Heim identifies several technical contexts in which virtualities have been digitally operationalized to attempt to create cyberspaces within the limitations of technologies available:

> Simulated 3D space on 2D monitors; interaction with electronic representations; immersion in hard- and software environments; the telepresence familiar from keyhole surgery (in which a video endoscope (a video-camera using a very small lense at the end of a fibreoptic cable) guides the use of surgical instruments to conduct an operation through a very small incision); 'full body immersion' in digital environments; and immersive computer-mediated communication networks allowing one or more users to interact in virtual space.
>
> (Heim, 1993: 110–116)

William Bogard argues that simulation has become one of the touchstones of the current era, 'hyped today in the market for things like global online services (simulated communities/ markets, information "highways"), genetic mapping and engineering (simulated bodies and body parts), expert systems (simulated knowledge), and virtual reality (simulated space and time, fantasy onscreen)' (Bogard, 1996: 14). He warns:

> Surveillance approaches something like an ecstatic form – where a single person wired into a computer can access millions of files anytime, anywhere, where wars are fought onscreen in satellite-fed electronic command and control centers . . . before deciding to fight them 'for real,' where parents (someday soon) will choose the genetic 'histories' of their children, who in turn will reflect less the biological differences of their mothers and fathers than the homogeneity of programmed norms of health and beauty. Such fanciful scenes invoke a futuristic landscape of surveillance without limits – everything visible in advance, everything transparent, sterilized and risk-free, nothing secret, absolute foreknowledge of events. But surveillance without limits is exactly what simulation is all about. Simulation, that is, is a way of satisfying a wish to see everything, and to see it in advance, therefore both as something present (or anticipated) and already over (past).
>
> (Bogard, 1996: 14–15)

Commentators in the first decade of the Internet – Arthur Kroker and Michael Weinstein, Vivian Sobchack and Ziauddin Sardar among many others – documented the sense of change and difference between online interactions dominated by virtuality and the face-to-face actuality of talk (still an ideal reference point for academics despite the widespread take-up of the telephone). They captured a sense of the phenomenological implications of cyberspace and 'focussed on the superficiality of the medium: its erasure of referential origins in favour of virtual presences, of spatial and semantic depth in favour of the shallow surface' (Cubitt, 1998: 144; see also Kroker and Weinstein,

1994; Sobchack, 1995; Sardar, 1996). Because cyberspace, or the digitally virtual, has been treated as ahistorical, located outside of longer trajectories of cultural and technological development, it has been made to appear as an awesomely magical yet violent rupture. If this has been experienced before, as the above example of the nineteenth-century panorama suggests, it is important to identify specifically what is different about digital virtualities, a question which requires that the virtual be understood in its relation to the concrete actualities of the everyday (Table 3.1).

VIRTUAL REALITY AND VIRTUAL ENVIRONMENTS

Digital, computer-based types of virtual reality (VR) have been developed in R&D labs since the early 1970s. VR is broadly defined as a computer-generated simulation or presentation of an environment in which the user experiences a sense of phenomenological presence or immersion in the environment (early definitions include those by Krieger, 1986; Benedikt, 1991; Biocca, 1992; Robinett, 1992; Pinsky, 1993). VR is now a popular term which describes an experience wrapped in a media-hyped idea and packaged as a set of applied technologies. VR 'draws together the world of technology and its ability to represent nature, with the broad and overlapping spheres of social relations and meaning' (Hillis, 1999: xv). Ken Hillis takes great care to distinguish the popular term *virtual reality* (VR) into what he calls 'virtual environments' (VEs) or digitally generated spaces and 'virtual technologies' (Table 3.2). Virtual reality systems may be divided into two clusters of technologies: simulation technologies and computer-mediated communication technologies. Virtual environments (VEs) are digital 'stage sets' and the available 'dramatis personae' (whether they be cartooned avatars, stylized bodies, Jurassic Park-style animations or talking flowerpots) in VR.

Virtual reality

VR environments extend the idea of cyberspace, a spatialized representation of digital domains and data in which users engage with each other primarily by interacting with data and messages (for a discussion of the ideal cyberspace see Novak, 1992: 225). Beginning with the widespread use of real-time chat (IRC, below) a sense of immediacy was possible because both users see each other's message as each person types. Sheridan proposed that three key elements of VR are:

- sensory information;
- control of relation of sensors to environment (i.e. ability to move);
- ability to modify the computer-generated environment.

(Sheridan, 1992: 121–122)

However, the social needs to be included. The most talked-about possibility of VR is interaction with others, not objects or environments (whose more simple qualities have ironically allowed VR to succeed but not as a communication medium (see below)). Thus the success of VR has been measured by 'the extent to which other beings also exist in the [virtual and real] world and appear to react to you' (Heeter, 1992: 262). This would be not only a new way of working with data but would allow new forms of social interaction. Access alone, for example, would grant entry to a new sensory world, and a privileged expansion of consciousness was imagined.

The essence of the history of VR is a series of attempts to actualize a paper published by Ivan Sutherland, the inventor of the first interactive graphical device 'Sketchpad'.[2] 'The ultimate display' proposed an immersive 3D graphical display (Sutherland, 1965). First developed as a 3D head-mounted display helmet, the objective was to link more closely the user's mind and computer. The modelling of chemical molecules, architectural projects and flight training were early adopters of various types of software which allowed 3D modelling and produced

Table 3.1 Chronology of Internet and virtual reality technologies

	Internet	*Computers*	*Virtual reality*
1930s			Cinerama, first flight simulators
1940	First remote connection to a calculator	First computers	
1946		ENIAC 1	
1951		First commercial computer: UNIVAC	3D films Hollywood
1954		Magnetic drum data storage	
1955		FORMATION OF ARPA	Remote stereo camera, head-mounted display
1960		PDP-1 for research use	
	Information Processing Techniques Office (IPTO)		
	Sutherland Head of IPTO		'Sketchpad' program
	Taylor Head of IPTO		First visions of digital virtual reality
1965	Davies (Cambridge) and Beran (RAND): packet switching		
	BBN awarded Arpanet contract		
	First Arpanet node		First VR machine (Sutherland)
	4 Arpanet nodes		Responsive environments (Krueger)
1970	Email first used		
	TCP/IP developed		
1974	Networks established in UK, France, Germany, Japan	First consumer computers	
1975	Computer Bulletin Board System (CBBS)		

	First BBS: COMMUNITREE		
	Usernet		
	First MUD		
1980	MINITEL (France)		
1981	Fidonet: Internet email protocol	Microsoft MS-DOS program	Dataglove invented
	CSNet	IBM PC	
	BITNET	Apple MacIntosh	Virtual Environment Workstation
1985		Microsoft Windows	
	Cleveland Free-Net		Full body suit
	CSNet & BITNET → CREN		
	Internet relay chat	Document scanning and	
1990	Archie, WAIS,	optical character recognition;	Battletech arcades (USA)
	Gopher programs	Corel Draw program	
1994	WWW invented (CERN)	Netscape Mosaic web browser program	
		Yahoo! Web portal launched	
1995	Broadcasting across the Web	Word for Windows program	VRML
		VRML sites: Activeworlds.com	Jaron Lanier
1996	London stock exchange:	Web-based video-conferencing	VRML MUDS
	'big bang' computerization	Ebay	Fakespace systems: The Cube
1997		Widespread availability of CD-writers	
1998		High-powered home computer graphics	Large-scale VR training simulations
2000	Online Music downloads (Napster)	Widespread use of digital cameras	Dot.com industry meltdown
2001	Enron collapse (virtual markets)	Microsoft anti-trust case re browsers	
2002	G3 mobile phones (Japan)	Ubiquitous computing	

Source: after Kitchen (1999: 27)

Table 3.2 Aspects of virtual reality

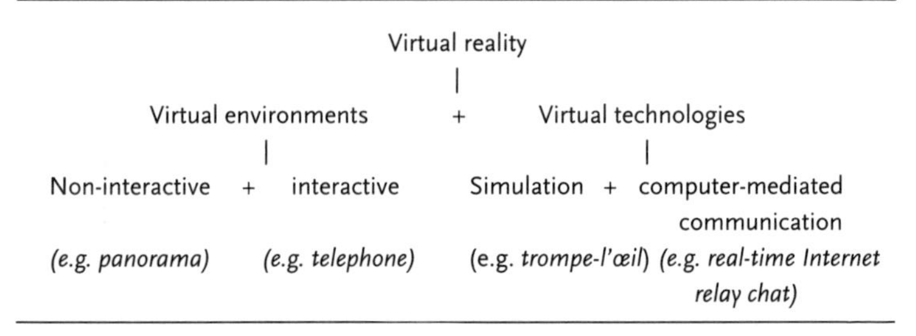

Virtual reality			
Virtual environments	+	Virtual technologies	
Non-interactive +	interactive	Simulation +	computer-mediated communication
(e.g. panorama)	*(e.g. telephone)*	(e.g. *trompe-l'œil*)	*(e.g. real-time Internet relay chat)*

images with limited interactivity, such as allowing the user to turn the image. In animated form this might give the illusion that the viewer was moving through the model being displayed on screen. More sophisticated systems extended the illusion of movement or added stereoscopic displays, 3D glasses or visors with devices to track eye movement and redraw the display.

Others such as Myron Krueger and researchers at MIT experimented with environments using wall-sized displays through which a user could navigate using a glove-like pointer (see the discussion of the Nintendo dataglove, Chapter 5). The problem with all of these systems was their unreliability, extreme cost, the need for extensive programming to create environments to be displayed, the lack of computing power to rapidly re-render changing graphics, and the lack of storage devices to allow digital photo images to be integrated seamlessly into digitally created visual scenarios. Applications were primarily military.

Although developed in the early 1980s for training and as sophisticated information interfaces for fighter pilots, by the 1990s some of the virtual reality simulation technology was commercialized using the growing power of home computers. Firms such as Nintendo and SEGA created goggles and gloves, while Sony and Microsoft implemented force-feedback (vibration) effects in handheld game controllers (see Chapter 5). Large-screen televisions and computer screens were the closest most got to a virtual environment.

Games of the late 1980s, in which users 'flew' chunkily drawn airplanes across landscapes rendered in polygons with only an occasional landmark, quickly became more realistic environments in which players could animate an avatar or character to explore exotic virtual environments and fight fantastical opponents (for example, the first shared virtual reality arcade game *Dactyl Nightmare*; see Kitchin, 1998: 49). The best-selling game based around the avatar Lara Crofts featured a kick-boxing, buxom archaeologist character – a male teenage fantasy of femininity as dangerous yet fascinating (see Wyatt *et al.*, 2000; Green and Adam, 2001), to be controlled by the male user yet independent in the plot lines and stories created around the character in promotions and movies.

Sex and mechanical dinosaurs are central to the history of VR. Fantasies of force-feedback body-suits and 'teledildonics' provided heterosexual visions of virtual sex. Even though these were not always explicitly discussed, nor were they the topic of the games sold (which included everything from team sports and individual combat to Formula-1 car-racing). The scenario of being able to give and receive sexual gratification via the Internet in some sort of personal virtual reality suit provided a hidden discursive unity to the efforts of young, mostly male, engineers engaged in speculative programming and experimental tinkering for the US Air Force, NASA or for telecommunications labs. Meanwhile cinematic special effects focused on animating dinosaurs for films such as *Jurassic Park* or creating walking, tyrannosaurus-like robots and 'battlemechs' which by the close of the century were a staple of children's toyboxes.[3]

Virtual environments

For computer programmers, it was a small step from databases to geographic information systems (GIS) which plotted data according to geographical coordinates in the form of maps. In addition, it was obvious to depict objects according to 3D spatial coordinates. But this is a relatively recent development, beginning from the 1980s. This late twentieth-century transformation of the idea of mapping was not transformed into a multimedia concept until recently. Hughes summarizes this idea of 'the virtual' tersely:

> The gist of the conception is that future computer technologies will allow users to become acting elements in a space engineered and defined by the technology; elements and spaces which need bear little relationship to how we understand our present embodiments and their spatial location.
>
> (Hughes *et al.*, 1998)

This space depicted in a VE need only follow the rules of everyday Euclidean space and geometry for the sake of providing familiarity for users. One's appearance could depend on one's mood; things that appear one way to one participant may be rendered to another as an entirely different object in their virtual space. One could imagine different levels of 'privilege' in these digital worlds – some might be able to see or do more than their data-poor neighbours in one shared virtual environment hosted by a computer in a third location. Others see digital virtualities as an extension of the drive to conquer and control. The positioning of VEs as 'in-between' various participants who share access opens up the possibility that they will be treated as liminal spaces betwixt and between not only people or locations (see Chapter 1). Digital virtualities offer themselves as *deterritorialized* spaces of escape from norms. Hence the allure of cyberspace as a haven for those who are otherwise labelled deviant or who feel the restriction of social and moral discipline too strongly.

Virtual environments are simulations characterized by four elements (after Cubitt, 1998):

- the primacy of navigation and movement;
- smoothness or unity of the digital environment, which includes a computer-generated character or avatar representing the user;

- a single 'point of view' (POV) which represents the user's position and outlook on to the VE;
- implied off-screen spaces.

Within VEs, navigation is identified by Cubitt as the primary structural device of contemporary virtual environments such as digital simulations. A second feature is a contrast to cinematic cuts between 'shots': smoothness.

> This viscous unity of the previously disparate [data] can be regarded as a function of the suppressed history of digital cinema – cross fades and virtual swoops from scene to scene suppress the montage effect of cinematic editing where one jumps from shot to shot. Morphing technology 'allows characters to melt into liquids or segue effortlessly from male to female or human to machine', promising, 'liquid identities in a liquid world'.
>
> (Sobchack, 1998, cited in Cubitt, 1998: 79)

However, most VEs still rely on the cinematic idea that the virtual space extends off-screen even though it can neither be seen nor accessed. Hence the popularity of game settings such as labyrinths, prisons, caves and interior chambers of pyramids and the like. These spatial frameworks efficiently spatialize a virtual environment, endowing it with the implicit sense of being an extensive environment. Furthermore, it supports the illusion and the trick of *metaxis* (see Chapter 1) by which the space between the screen and the eye is filled out into an extension of the digitally virtual VE, even though it is displayed in only two dimensions, captive to the surface materiality of the display screen.

The moving camera, of which live television was the ultimate example, anticipates the 'point of view' (POV) which roams the VE. As a surrogate set of eyes, the vantage point displayed allows the digitally virtual space to be seen from various different angles. However, as Hayles points out, the POV is associated with the user's position and with 'me' – it represents subjectivity within the computer-generated scene (Hayles, 1993). Thus a mobile POV, generated by a moving camera, may be interpreted as virtual movement, as travel by the user within the VE (Shields, 1996: 87). 'The camera in cinema, like . . . the panorama and the diorama . . . mobilizes the audience across the gulf that opens now between static spectator and mobile spectacle' (Cubitt, 1998: 78). Television offers a sort of mastery of this space, like Jules Verne's universal porthole: 'allowing the viewer to select any current activity on the face of the planet to look in on. The visual media of the moving image embraced the prospect of vision as unlimited travel' (Cubitt, 1998: 78; see also Friedberg, 1993: 109–148).

Simulation technologies

As noted in earlier chapters, simulation technologies range from simple *trompe-l'œil* tricks of perspective to elaborate moving 3D environments. The history of VR (above) is usually described in terms of the technical advances in simulation technologies. These include interfaces such as high-speed software for rendering virtual environments graphically, and stereoscopic display goggles or head-mounted displays which give a sense of three-dimensional vision. Force-feedback and other haptic devices add tactile markers and sensations to various aspects of a display but usually concentrate on achieving a feedback effect through a hand-operated controller (as in a video-game controller, joystick or a mouse).

While they have been widely showcased in the media, most simulation technologies remain cumbersome and beyond the average consumer. However, in industrial and medical applications VR plays a prominent role in training simulators for a range of occupations, which require precise operation or actions over long stretches. Examples of where these VR training simulators have proved useful include simulated training of airline pilots, simulated heavy construction crane operation training and simulated operation situations for surgeons (see Chapter 6). Attempts to simulate the experience of flight for training pilots go back to the 1930s. These do not depend on an extensive infrastructure and high-speed Internet access. Inhospitable environments can be entered with robot vehicles controlled using different types of virtual reality technologies to display the environment and relay an operator's actions, such as picking up objects or using types of datagloves. These activities include the storage and maintenance of radioactive materials, deep-sea-diving operations, assembly of satellites in space and are expected to include keyhole surgery and so on. Few studies of the diffusion path by which VR technology may be taken up and find more widespread applications have been done (but see Valente, 1995). Above all it is the military that has invested most heavily in virtual reality interfaces and simulation systems such as head-mounted displays (Pimentel, 1993: 30ff.).

Computer-mediated communication technologies

Virtual environments are a type of interactive communication medium which changes our understanding of our embodied nature and the limits of our everyday world. By aspiring to be 'as good as' a face-to-face meeting, they challenge what Dede Boden called the 'compulsion to proximity' and our notions of co-presence.

Computer-mediated communication (CMC) may include software and hardware for dial-up networking over telephone lines but high-speed optical systems such as the most recent high-speed cables which make up the backbone infrastructure of the Internet are required for adequate speed or bandwidth to achieve a sense of real-time interaction. Not only redrawing but transmitting the changing positions of participants and objects in virtual environments continues to be a challenge. Shared virtual environments in which participants are at a distance and meet in a simulated environment depend on both of these clusters of technologies. Thus 'telepresence', a sense of presence *to others* via the mediation of technology, is a defining mark of shared virtual environments.

The Internet itself was originally a brainchild of the US military, which sought to create a decentralized computer network, ARPANET, which could function despite the loss of any single node in the network (see Table 3.1; for a thorough history see Kitchin, 1998; Abbate, 1999). Each computer was linked to a number of others, and data passed from computer to computer until it reached its addressee. This means that information often takes a round-about route, resulting in delays. This is the Achilles' heel of the system. Downey argues that the key process of the Internet is digital convergence, a process which is not only technological, but social, commercial, legal and political. Information from previous communication networks is recoded, stored and transmitted digitally. Convergence draws together the 'technologies, institutions, commodities and labourers of three pre-existing networks' (Downey, 2001: 212):

- packet-switching computer networks such as the old ARPANET and today's Internet backbone;
- telephone networks;
- wireless radio and television broadcast networks.

The result is to decentralize the technical control centralized in a previously single system and to establish a shared protocol among all component systems (Edwards, 1998) such as the single Internet Protocol or 'IP address', a unique twelve-digit number which identifies any Internet-accessible computer worldwide, but which has not yet prevailed over the more ubiquitous household and business telephone number.

Mated with the graphical interface of the World Wide Web and browsers capable of 'reading' a simple hypertext markup language (HTML) and displaying the results on one's computer screen, the textual world of email and the exchange of typed messages between a number of interacting participants in a 'multi-user dungeon' (MUD) or in the Internet relay chat channels (IRC) of the early Internet blossomed within less than a decade into animated sound and pictures of 'web pages'. The number of Internet domains (or named sets of sites under one owner) rose from around 15 million in January 1995 to around 125 million in January 2001 (Internet Software Consortium (2001) figures). However, the data are difficult to assess. There are significant fluctuations in users depending on the time of year. This suggests that student and workplace use is an important aspect of the Internet in general and likely to be significant in the use of forms of digital virtuality (see Chapter 6, and Table 3.3).

Table 3.3 Number of hosts advertised in Domain Name Survey

Year	*January*	*July*
2001	109,574,429	125,888,197
2000	72,398, 092	93,047,785
1999	43,230,000	56,218,000
1998	29,670,000	36,739,000
1997	16,146,000	19,540,000
1996	9,427,000	12,881,000
1995	4,852,000	6,642,000

Source: Internet Software Consortium Internet Domain Survey, July 2001

APPLICATIONS OF VIRTUAL REALITY

'VR *is* a novel form of training ground on and in which users learn to overcome what would have been until recently resistance to the incoherent proposal that they might occupy the space of an image' (Hillis, 1999). Furthermore, it is a training ground not only in particular ways of seeing but in ways of imagining fictional, distant and alternate realities. VEs offer the possibility of digitally restaging historical events and injecting users into the scene, *Zelig*-like, for educational purposes. They offer a host of possible ways of allowing one to be present in distant or hostile environments. There is no obligation for virtual reality

(VR) to conform to physical reality; indeed, the opportunity to create alternative worlds – fictions – is one actual appeal of virtual reality technologies (e.g. see the simulated but fictional world 'Mars' with its homesteads and online community hosted by Active Worlds (*www.activeworlds.com*)). On the one hand, current simulators may advertise the correspondence of their virtual environment (VE) with a real environment. For example, commercial golf simulators project a floor-to-ceiling video image of a choice of famous golf courses on a movie screen while radar and impact sensors detect the speed and direction of golf balls hit against the screen. The trajectory of the ball is plotted (based on user-selected factors such as cross-winds) and displayed flying off into the scene from the point at which the actual golf ball hit the screen, and the view up the course is displayed from where the ball has 'landed'. Needless to say, one cannot play backwards and one cannot pursue the ball off the course; neither does one ever meet or see any other players on the course.

On the other hand, many role-playing simulators use the same technology but scramble up references to real places. For example, Hillis discusses the US Army Research Labs and Division Incorporated's combat simulator in which an apparently historic English village might be set in a desert surrounding. As a 'grab-bag' of cultural borrowings,

> if landscape is understood as the visual aspects of a place, then a strong sense of landscape is achieved, albeit one that is highly geometric in execution. However this visual sense of looking into a scene constructed according to laws of geometry and perspective is not the same as how we see the real world . . . the meeting of the English village with the middle of nowhere begs the question correspondence with what, with where? Correspondence in VR need not be with any real place on the earth, but rather with imaginary places and circumstances made to seem real enough by an appeal to aspects of visual perception responding to texture . . . of surfaces and so forth.
>
> (Hillis, 1999: xxviii–xxix)

In effect, these are not only environments that are produced via programming algorithms for rendering 3D settings on 2D video displays, nor simply representations that mimic the common syntax of spatial perception and action in the world (for example, by setting the speed at which a character or the player's point of view can move through the VE). They are 'spaces of representation' (Lefebvre, 1981) that embody particular ideologies and fantasies even in the way in which conventional objects are depicted, and in the visual rendering of characters and of the landscape.

With the restructuring of the computer industry in 2001 and shrinking expenditure on technology by businesses, the two remaining 'killer apps' for ambitious, large-scale virtual environments are design and visualization, on the one hand, and training, on the other. Some firms remain focused on military training, such as the examples given above. The Institute for Creative Technologies at the University of Southern California is attempting to create a Holodeck-like virtual environment along with teams from other American universities. This would be filled with characters who will interact with trainees. One scenario this could be used to simulate would be peace-keeping tasks requiring good relations with civilians (Scott, 2002: SP7).

Architectural 'walk-throughs' or the ability to display and walk around a proposed automobile design are well worth the expense of large-scale immersive environments. GM describes it as a 'major change in the way we develop design' (J. Attard, General Motors, quoted in Scott, 2002: SP6). The reigning 'brand' in the world of virtual environments is 'The Cave', a convincing 'illusion machine' that consists of a 3m × 3m × 3m room-sized simulator with wall-sized video data projections and head-mounted eye-tracking goggles which allow the scene to be redrawn around the user as they look around. In part because one can move about, it allows one to suspend disbelief so effectively that one car company executive is said to have put his coffee cup into the virtual coffee-holder depicted in a car interior in The Cave (Scott, 2002: SP6). The Cave system is manufactured by

Fakespace Systems in Kitchener, Ontario, and is installed in many places from the Virtual Museum in Linz, Austria, to the New York Stock Exchange, the Tokyo Opera Centre, and at many labs and manufacturing corporations. At around US$1 million, they are inexpensive compared to earlier large-scale simulators.

While the New York Stock Exchange (NYSE) still uses people instead of computers to trade, it monitors the estimated US$20 billion in trades each day via a 'computer world that makes you feel you are going into a three-dimensional space – with the advantage of being able to zoom from an overview to a close-up in an instant'. The NYSE is a 'machine for information exchange' . . . now simply re-created in a different space (H. Rashid, Asymptote Design, quoted in Scott, 2002: SP7):

> Although the virtual NYSE includes live footage from the floor, its main architecture doesn't look like the real thing. [Designers] Rashid and Couture rebuffed suggestions to make the walls look like marble. 'We looked at them and said, what does that have to do with information? . . . Why not put data in the walls.'

VR that mimics the concrete world while remaining closed in upon itself is pointless, myopic and even blind. 'How did . . . [they] turn into a hollow, emptied-out model . . . The majority of the stuff [VR] [out there now] is kind of illustrative. It's the Norman Rockwell syndrome' (H. Rashid, Asymptote Design, quoted in Scott, 2002: SP7).

What is missing in virtual models of places one can visit in the non-virtual world is not just wind and sun but memory. A virtual tour of the Pyramids is more likely than not to include only the Pyramids: no chance encounters with other visitors, guides, no risk of discovering the current conditions of life in Giza from encounters with guides, touts and hustlers. The difference between the NYSE project and most other simulations is that it includes information and images flowing constantly into The Cave.

Hopes for consumer applications of virtual reality technologies centred around the videotape and the cluster of technologies such as colour TVs and monitors, CD-Rom drives, cable broadcast networks, digital video or webcams, and the telephone. The Mattel PowerGlove™ was marketed as a dataglove for the Nintendo videotape system at a cost of US$89 (1989–1991). However, these all have technical limitations which have foiled most attempts to bring them into what would be a new arrangement to support VR affordably. This includes using the local cable TV networks for two-way communication. Even without moving to optical cable, co-axial cable TV wiring affords more bandwidth than the twisted pairs of copper telephone wire. By 2000, consumer computers for home and business use were powerful enough to run the interface, and CMC via cable TV was offered, but the Internet remains too slow even for the reception of uninterrupted audiovisual content, which must be buffered (that is, a minute or more of data are downloaded before being played, in anticipation of short interruptions or lags during the rest of the downloading process, which eat into the buffer of material). The problem is compounded when attempts are being made to both receive and send audiovisual communications (not to mention movement and spatial orientation data) in real time. Any 'crutch' such as buffers would create an intolerable lag for the participants – an exaggerated form of the delay one experiences on long-distance telephone calls if they are being relayed by satellite.

COMPUTER AS FILTER

By digitizing and displaying information on video screens, the computer becomes an important filter through which the real is experienced. Computer-based media have generally been considered in terms of how they encode 'reality', understood as the concrete. However, as argued earlier, the virtual has its own autonomy from the concrete. As rendered by computer and network software and hardware, digital virtuality is dependent on the technology: fidelity, resolution, bandwidth and the type of interface are the parameters of virtual environments created by computers. Furthermore, computer software invites interaction by users and manipulation of the material. This makes it quite a different medium from telecommunications media such as the telephone and television despite its network similarities with the phone (cf. Nye, 1997; Standage, 1998). The *matériel* of the real becomes *material* which is edited, transformed and archived in new formats which leave the moment of digitization far behind. Virtual environments are worlds of light. VR goggles etch the changing image of the virtual environment stereoscopically on each retina to give the illusion of depth. Sound and tactility are poor cousins to the visual sense, supporting or confirming it. While a type of chemical printer which releases smells was invented at the end of the 1990s, its lack of utility doomed it to novelty status. Because of the industrial and pragmatic nature of current VR applications, smell is a distracting and even unwelcome addition. The objective is to remove the user, the operator, from the actual world to allow action at a distance from the 'convenience' of a terminal.

Movement and gesture are also rendered visually – the floating glove or pointing hand which often represents the user's position and orientation. Rather than a 'haptic' VR, the effect is very similar to an automobilist's experience of the landscape whizzing by outside of closed car windows.

> Sounds in VE are related to the spatial relationship between user and icons. They are always the same for any one situation. . . . In this, digital sounds in a VE operate as "aural icons". Moreover, as audio theorist Steve Jones argues, the very jargon of VR excludes the aural (1993: 239), and the creation of VR "can be understood as part of the ongoing technological visualization and deauralization of space".
>
> (Jones, 1993: 246, cited in Hillis, 1999: xxii)

But here too there is debate. Coyne comments that

> McLuhan exposes the two great epochs of the senses: the aural sense and the visual. The sense of hearing is immediate and

> unitary. It is the sense of preliterate, tribal humanities, for whom there was a unity with nature and a lack of differentiation. The epoch of the visual sense came with the invention of writing and manuscript culture and pertains to distance, objectivity, classification, language and the symbolic order. The electronic age sees a return to the aural sense . . . in conflict with the symbolic order (the visual). [The aural is] . . . a kind of distributed reductionism that seeks the unity of all things in information, to be contrasted with Descartes's centralized reductionism, with the homunculus, the home of the soul.
>
> (Coyne, 1999: 232)

The actual *matériel* may thus be forgotten, lost or supplanted; hybrid images such as a film starlette's head on a porn star's body posted up on the web; even becoming the source for later content in which the now-digital material is further cropped, edited and changed.

Computers allow visual aids to be created which respond in real time to changes and patterns of information. For example, an 'up arrow' on screen indicates the increasing value of a stock-exchange index or the value of a currency. A digital model of the human skeleton can be displayed from any vantage point including from a point of view within the rib cage or can be travelled through cinematically and interactively (cf. software such as 'The Interactive Skeleton': Routledge/Primal Pictures). Icons and animations make abstract data tangible and informative by rendering it into the vocabulary and context of the virtual and by bringing it into relation with other information.

The debate rages as to what the impact of the digitally virtual will be; however, evidence of the virtual in literature and in social ritual (as argued earlier) suggests that this is a long-held human capacity for imagination and a perceptual flair for filling in the gaps and fleshing out visual images. Popular film and boosters of virtual reality technologies such as head-mounted displays suggested that we might soon be able to upload our consciousness into a computer system, interacting with the world through digitally activated devices (cf. the film *Lawnmower Man*, or Gibson's *Neuromancer*, 1984). Despite claims that virtual reality technologies offered the opportunity to leave the body behind and immerse ourselves in a world of pure simulations, leaving 'the meat' behind, virtual reality and cyberspace do not signal a shift in human nature or a new step in human development.

As proposed by a number of prophets of virtuality, technology promised an experience of oneness with a world of data, meetings with distant friends in simulated environments, and tactile experience. Breaking into the mainstream media in 1987, VR was touted as an entertainment technology (Lanier, 1992). Suspended weightless in bodysuits covered with movement sensors and actuators which would convey tactile sensations and pressures directly to the skin, an interactive massage could be digitized and transmitted to another. Imagining form-fitting devices, a type of cybersex was envisioned – 'teledildonics' was born. These fantasies remain mere techno erotica; imagined, not functioning prostheses, this 'digital virtualism' has found a home at work and practical deployment in its 'small' form as simulated 3D space on the flat displays of computer monitors. Animation, not immersion, has been the major application: in gaming software, in the real-time graphs depicting financial exchanges and in the cinema.

Still, VR 'theatres' which competed with cinema have been realized only in the form of art gallery installations such as Char Davies' *Osmose* (Davies, 1995). Sega introduced a headset with stereoscopic display goggles and earphones in autumn 1993 for around US$150. However, their use did not become widespread. For most video-game players, a more direct sense of participation in the on-screen milieu of computer-generated labyrinths, shoot-em up Wild West towns, the endless plumbing of 'Super-Mario'-type games and the landscapes over which simulated aircraft were flown, required more detailed graphics first. Far more could be achieved by the commitment of imagination instead of cash. For dedicated gamers who were mostly young, the boredom of our cityscapes, long car rides and waits for public transport favoured the escapism of the simple worlds of computer-generated games. The portability of the handheld, battery-powered 'GameBoy' won hands down over the cumbersome head-mounted display. A further problem with any system which covered the eyes was that it tended to limit the user to online immersion. Far from leaving 'the meat' behind, any observer of players would have noted the often social importance of the games, with friends playing as well or gathered looking over the player's shoulder peering at the screen. Without the ability to show off one's exploits, the computer game loses its social appeal.

THE AUTONOMY OF THE VIRTUAL

Earlier chapters demonstrate the autonomy of the virtual from the concrete, but also point out the choreographed interweaving of the abstract, virtual and concrete in everyday language and action. What appears most interesting is the movement between these ontological categories, the conjuring of elements out of one category into another via symbol, memory, metaphor, the concretizing quality of action, calculation and prediction. In the case of the digitally virtual, there has been much debate concerning the autonomy of the Internet as a separate realm or sphere in which interactions could be conducted outside of the legal structures and social norms of existing states (online and collected in print in Ludlow, 2001). In response to the Clinton administration's attempt to regulate speech in all media, a law that presumed to regulate all users internationally, John Perry Barlow widely emailed his 'Declaration of the Independence of Cyberspace'. Part of this reads:

> Cyberspace consists of transactions, relationships, and thought itself, arrayed like a standing wave in the web of our communications. Ours is a world that is both everywhere and nowhere, but it is not where bodies live.

> We are creating a world that all may enter without privilege or prejudice accorded by race, economic power, military force, or station of birth.
>
> We are creating a world where anyone, anywhere may express his or her beliefs, no matter how singular, without fear of being coerced into silence or conformity.
>
> Your legal concerns of property, expression, identity, movement, and context do not apply to us. They are based on matter. There is no matter here.
>
> (Barlow, 1996)

In response, writing in the online journal *C-Theory*, one commentator pointed out the rootedness of all digital virtualities in a digital materiality: the corporate servers and network cables fixed in place:

> In sci-fi novels, cyberspace has often been poetically described as a 'consensual hallucination.' Yet in reality, the construction of the infobahn is an intensely physical act. It is flesh and blood of workers. . . . It is obviously a fantasy to believe that cyberspace can ever be separated from the societies – and state – within which these people spend their lives. Barlow's 'Declaration of the Independence of Cyberspace' therefore cannot be treated as a serious response to the threat to civil liberties on the Net posed by the Christian fundamentalists. . . . Instead, it is a symptom of the intense ideological crisis now facing the advocates of free-market libertarianism within the online community. . . . Crucially the lifting of restrictions on market competition hasn't advanced the cause of the freedom of expression at all. On the contrary, the privatization of cyberspace seems to be taking place alongside the introduction of heavy censorship. Unable to explain this phenomenon within the confines of the Californian Ideology, Barlow has decided to escape into neoliberal hyper-reality rather than face the contradictions of really existing capitalism.
>
> (Barbrook, 1996)

Although policing is never 100 per cent successful, by restricting Internet access to only a few, closing down servers when necessary, censoring telecommunications and policing individuals who use the Internet, governments such as China have effectively strangled the development of the Internet and its use by their citizens (see Chapter 4). Meanwhile an informal market beyond the reach of taxes operates in cyberspace up to the point at which material objects cross customs boundaries where taxes can be levied (cf. Ludlow, 2001: 5). Disputes such as trademark and intellectual property, which cross legal boundaries, can emerge in cyberspace. These suggest the emergence of a virtual jurisdiction (see Schachter (2001: 85–98) on the seminal legal case involving America Online):

> Global computer-based communications cut across territorial borders, creating a new realm of human activity and undermining the feasibility – and legitimacy – of applying laws based on geographic boundaries. While these electronic communications play havoc with geographic boundaries, a new boundary – made up of the screens and passwords that separate the virtual world from the 'real world' of atoms – emerges. This new boundary defines a distinct cyberspace that needs and can create new law and legal institutions on its own.
>
> (Johnson, 1996)

Others feared that cyberspace is being constructed precisely as a 'Temporary Autonomous Zone' (Bey, 1985), to avoid the responsibilities of everyday life: an 'island in the Net' outside of its duties and jurisdiction. In its first decade cyberspace was widely regarded as a liminal zone outside of the everyday. Jodi Dean writing in 1998 pointed out that cyberspace seemed to solicit and even produce millennial cults such as *Heaven's Gate* and supported the dissemination of stories of abductions by aliens and UFO encounters which attacked official statements on the nature of mysterious events (Dean, 1998). Recall from Chapter 1 that this is a battle to allocate events to one or another status in the tetrology of the virtual and actual. Are the experiences of abduction recovered under hypnosis been merely figments of the imagination (i.e. abstractions), virtualized via rituals of recollection and recovery, and actualized by public retelling online and on lurid daytime television chat shows? Are these 'real' or 'false memories' (i.e. virtual or abstract)? In cyberspace, on daytime television and even in learned essays, the carnivalesque tinted the 1990s. Groups increasingly challenged 'official reality'. The new telecommunications infrastructure, crude display technologies and widening access to the Web allowed critiques of ongoing trials despite media bans by courts (e.g. the trials of serial murderers such as Homolka and Bernardo), and the presentation of detailed arguments about the cause of air crashes, the flaws in aircraft construction, and possible military or terrorist attacks.

Virtual technologies may be accused of incorporating 'an allegiance to the idea that all of reality is a social and linguistic construction' (Simpson, 1995: 142); to 'the demiurgic desire to be the origin of the "real"' (Simpson, 1995: 140, quoted in Hillis, 1999: xxix). Taking aim at the equally individualistic philosophy espoused by the editors of *Wired* magazine, Purdy writes: 'A few people, mostly college students, have largely withdrawn from their embodied lives to participate in virtual communities. . . . By entering these realms, their programmers reproduce the "old theme" of "the god who lowered himself into his own world." Kelly [the editor] identifies this theme with Jesus, but one wonders if Narcissus is not a more appropriate touchstone' (Purdy, 1998). None the less, 'Users of non-immersive telecommunications, such as IRC and other chat-room environments accessed via internet . . . experience a "feeling of place", believe they are "in" something, are some "where"' (Dery, 1993: 565) which continues to raise questions of the material locatedness of the digitally virtual. Hillis summarizes this literature, asking whether or not there can be any meaningful 'geography' of cyberspace worth studying (Hillis, 1999: xxx). The challenge is to link together the material locatedness of servers, users and actual places with the virtuality of cyberspace and the virtual components of everyday life, our memories, our dreams and our metaxic ability to integrate and slide between the concrete and the virtual.

This unity is often sacrificed, a hallmark of both promoters and boosters of the Internet and VR and the alternately awed, panicked and overstated first-generation critics of cyberspace. Mark Dery identifies a strain of utopian disdain for the concrete in the rhetoric of prominent essayist Nicholas Negroponte: 'Troubling thoughts of social ills such as crime and unemployment and homelessness rarely crease the Negroponte brow. In fact, he's strangely uninterested in social *anything*, from neighborhood life to national politics' (Dery, 2001: 391). Dery writes off digital attempts to actualize the virtual while pointing to the cynical capture of the virtual for the purposes of profit, which goes beyond the publishing empires from Victorian fiction onward. Negroponte and his MIT Media Lab are:

> an assembly line for vapourware, technologies that exist only as consensual hallucinations in the mass mind. The quintessential piece of vapourware is virtual reality, a technology that was obsolete before it ever really existed. Collapsing under the weight of the impossible expectations shovelled on top of it by cyberhypesters, VR was a victim of overexposure. . . . Obviously, VR exists in literal fact, but the crude, polygonal state of the art falls far short of the disembodied ecstasies evoked by Jaron Lanier and William Gibson. Like VR in its early 1990s, mass-media incarnation, commodities of the future will be consumed as concepts only, living out their fifteen-minute life cycles in the *vivarium of the mass media*.
>
> (Dery, 2001: 396–397)

Rather than the community-based experiments in virtual utopia of Net and MUD pioneers, the virtual, especially in its digital incarnation, may well have become a mass-media product, sold like a ticket to the theatre by huge integrated media conglomerates such as Sony, AOL/TimeWarner in North America and Japan or Vivendi in Europe. For those with high-speed wide broadband access; patented subscriber VEs; for those with slower dial-up modems, $50 video-games with online, multiplayer modes (see Chapter 5); for those with no access, Hollywood films and videos such as *The Matrix* that hype a wired future of VEs for all at vanishing point set off in the future.

In the context of the digital, Hayles provides a more far-reaching definition of virtuality as 'the cultural perception that material objects are interpenetrated by information patterns' (Hayles, 2000: 69). One looks past or beneath the concrete form for a genetic code or the software that gives form. It is not merely psychological, but is a mind-set that is manifest in advanced technologies. Embedded at the heart of genomic and digital technologies is the division of software and hardware; information and materiality, structuring code and surface appearance (Hayles, 2000: 73).

The 'first-generation theorists' of cyberspace mapped the virtuality of digital communications media on to a dichotomy of spirit and matter, with matter fixed firmly in the familiar world of the body. But they devoted less attention to the divide between the liminal and everyday life, between two modes of sociality. The liminal offers itself as the exceptional, a fluid space of potential – of virtualities – which cannot be inhabited. It does not offer the stability and robustness required for dwelling. The consolidation of culture and the reproduction of society requires the fixing of social norms as enduring structures which outlast a single generation so that they may be handed down. Thus the rapid pace and fluid instability of digital simulations pose a challenge to attempts to fix and institutionalize culture, to develop and propagate norms of behaviour which are seen as legitimate and to stabilize values by embedding them in concrete forms such as monuments, buildings and cities. The digital virtual offers only a technique of simulation and memory which is being used to model and anticipate the future.

SUMMARY

This chapter has considered the rise of simulation soft
hardware as a digital form of the virtual. From the paint
circular panoramas of the 1800s to immersive virtual reality and digital renderings of environmental role-playing games, there is a long history of virtual environments. Central to the recent history is the rise of sophisticated graphic display hardware and software, complex geographical information systems (GIS), and military and entertainment technologies such as video- and computer-games. Digital virtualities are synonymous with simulation, a process which was argued to be a liminoid genre, both standing outside of materiality of everyday and embodied life, and also lying between different players' computers. They are variously outside of or beyond, betwixt and between the environments or concrete locales occupied by the human user.

First generation theorists of cyberspace focus on the dichotomy of information and the body, spirit and matter. This false dichotomy masks the manner in which the two become intertwined. Liminal, digital virtualities are used as a 'technical fix' for concrete obstacles and problems in everyday life. Virtual environments are digitally created spaces of escape. Their status is ambiguous – outside of the material world, yet also in-between players' computers and dependent on telecommunications networks. The digitally virtual is thus embedded in the ongoing life of the concrete, standing in close relationship as a fantastic escape attempt, a simulation of possible events, or a rendering which is used as the basis of decisions on and actions in the material world. Rather than simply dismissed as 'vapourware', the digitally virtual is an important extension of notions of reality and the context of action.

However, the computer becomes a filter: digitized representations and records may become the focus of interaction and reference rather than the original object or event. Cyberspace provides an example of the binding of the virtual, concrete, abstract and probable into a complex whole which has measurable impacts on everyday life.

4

VIRTUAL AFRICA

Where is virtual Africa? In the heart of some deepest, darkest continent? Where could one go to escape the cyberspaces, the electronic chat rooms and virtual communities in the webs of telecommunications networks? These are now easily available in rickety cyber cafés at the high-tide-mark of the beaches of Third World tourist destinations. Ko Samui; Gold Coast; Los Islas dos Angeles; Ismir? Whatever the tags and stickers on our suitcases, it is difficult to escape the reach of the Earth's telecommunications girdle. The 'outside' is disappearing.

Africa, however, remains largely 'over the horizon' of the telecommunications systems. As a continent comparatively free from the digital virtual except for those able to pay the exorbitant costs of time in Internet cafés, it is oddly both excluded and out of the reach of global media. This chapter provides some insights into the relationship between the virtual and the global, using the counter-example of Africa as an illustration. In this chapter we will consider:

- globalization and virtual diasporic communities;
- digital virtuality and globalization;
- virtual Africa and the digital divide.

GLOBALIZATION

'As electrically contracted, the globe is no more than a village', wrote Marshall McLuhan in 1964. For many breathless observers of the scale and power of telecommunications media and corporations, Marshall McLuhan's notion of the 'global village' (McLuhan, 1964: 5) became fashionable again. During the 1980s,

> It was seen as the perfect expression of the new era. . . . In the 1980s, the financial system had migrated onto computer and communications networks, satellite and cable links that spanned the globe, capable of carrying data and voice, creating the conditions that produced the October 1987 stock market crash, where a fall in the New York stock exchange precipitated a collapse in prices that tripped off markets around the world within hours.
>
> (Woolley, 1993: 124)

If this wasn't 'globalization', what was? Globalization may be 'the' concept by which we name the current moment and the transition into the third millennium AD (for discussions see King, 1990, 1991; Appadurai, 1996). Most social scientists writing in European languages seem to accept that such a process is underway. Where there is debate, it concerns the impact of globalization and whether or not it is fundamentally new or rather a new wave of a process which has been ongoing for the past 400 years of European expansion, Westernization and what was called 'modernization'. The academic recognition of 'globalization' is signalled by the argument and collation of 1980s discussions of trade and of the dissemination of notions of Western civility and civil society in Roland Robertson's book *Globalization* (1992; see also Robertson and Lechiner, 1985); 'Globalization as a concept refers both to the compression of the world and the intensification of consciousness of the world as a whole' (Robertson, 1992: 8).

In the process of globalization, specific localities are argued to be linked together more closely than ever before, becoming interdependent (Keohane and Nye, 2000). Local happenings are shaped by distant events and decisions in a 'lateral extension of social contexts across time and space' (Giddens, 1990: 64). By forcing agents to make explicit decisions about which elements of the local economy and culture will be maintained and what aspects of global amenities and values will be adopted, globalization implies reflexivity and localization. Albrow summarizes globalization as 'all those processes by which the peoples of the world are incorporated into a single world society, global society' (Albrow (1990) in Pieterse, 1993). One introductory text defines globalization as:

> A social process in which the constraints of geography on economic, political, social and cultural arrangements recede, in which people become increasingly aware that they are receding and in which people act accordingly.
>
> (Waters, 2001: 5)

Although some suggest that globalization is only a one-way process of Westernization and modernization, Pieterse (1993), joined by Appadurai (1996) and Hermans and Kempen (1998), argue against reductionist views of globalization as Westernization. The former analysis is criticized for its reliance on a dichotomous appreciation of intercultural relations (Hermans and Kempen, 1998). The West is consistently opposed to the 'Rest' (Orientalism), and cultures are presented in binary oppositions, relying on internal homogeneity and external distinctiveness. There is no room for movement, exchange and deterritorialization (Cappeliez, 2001). Perhaps this is best understood as two successive 'waves' of globalization:

> The first wave of globalization – whether in economics or in media – witnessed vertical control from international centres, as witnessed for example by the rise of media giants such as CNN

> and MTV. But in more recent waves, a process of relocalization is occurring, as corporations seek to maximize their market share by shaping their products for local conditions. Thus, while CNN and MTV originally broadcast around the world in English, they are now producing editions in Hindi, Spanish and other languages in order to compete with other international and regional media outlets.
>
> (Warschauer, 2000: 156)

In the rebound effect of globalization, pirated copies of Western music and software have ensured the dissemination of Western pop, but eroded the profitability of Western media corporations. The 'second wave', late twentieth-century theories of globalization suggest the relativization of at least parts of local cultures and the displacement of the effectivity and integrity of the nation-state both as a polity and as an 'imagined community'. However, the difficulty of the concept 'globalization' is that it is approached via abstractions. The concept itself designates a process by which the concrete and local comes to be understood to be related to distant virtualities. Notions of locality are in a sense spatially 'warped' as they become entangled with the far-off and absent. They take on not only a variable scale but become 'over-dimensionally' extended. The local becomes less a matter of bounded, material place, and becomes virtualized – real but not so much tangible as a matter of essence.

In this process of 'hybridization', the fixed quality of localities is challenged by the distantiated, extended networks of diasporic communities whose localities are no longer proximate neighbourhoods but extensive family, alimentary and discursive exchanges. This 'glocalization' is carried along by specialty video rental stores and satellite telecommunications firms as much as ethnic grocers. No longer is the local a sign of withdrawal from the global but an essential node concretizing a rhizomatic network in which the global is only ever abstract and the local is an entanglement of the virtual and material (Table 4.1).

Table 4.1 Internet users 2001

Africa	4.15 million
Middle East	4.65 million
Latin America	25.33 million
Canada & USA	180.68 million
Asia/Pacific	143.99 million
Europe	154.63 million
World total	*513.41 million*

Various sources. Methodology compiled by Nua Internet Survey (August 2001)

DIGITAL VIRTUALITY AND GLOBALIZATION

> It was the excitement of being part of this world that stoked up the computing community's interest in virtual reality. Could this be where the denizens of the global village truly belonged. Could this be a *new* reality?
>
> (Woolley, 1993: 125)

The Internet is one part of the 'thickening' of global telecommunication connections. But in 1999, 1 per cent of the population in only 23 countries had computers permanently connected to the Internet (i.e. hosts – see United Nations Development Program, 1999): 'Although internet-related business firms often like to present figures indicating that internet use is doubling every few years, these figures usually refer to use only in developed parts of the world or, in the less developed nations, to a doubling of use from an extremely low base (e.g. from 0.1 per cent of the population to 0.2 per cent). For most people and countries of the world, becoming a significant player . . . remains far in the future' (Franda, 2002: 10).

The Web promised a global community of minds – a '*consensual* hallucination' in the words of the novelist William Gibson (Gibson, 1984: 67, italics added). Cyberspace is the idea of a virtual consensual community of belief. However,

> Insofar as it cannot – and does not wish to – recognize the historic material bases of community, 'virtual community' constitutes an increasingly shallow theoretical understanding in the midst of this redistributive communicative fragmentation. Instead, the analytic task becomes one of understanding the operation of the internet as a system of languages, where economic and class privileges accrue with gross differentials in the midst of racial and ethnic geographies.
>
> (Lockard, 2000: 177)

Web-based video conferencing (let alone virtual reality) remains difficult to sustain between even the two most 'wired countries', Canada and the USA, and much investment has been made to broadband 'backbones' in many countries. In 2000, analysts claimed that:

> Recent analysis indicates that the number of non-English websites is growing rapidly and that many of the more newly active internet newsgroups (e.g., soc.culture.vietnamese) extensively use the national language. . . . Indeed by one account the proportion of English in computer-based communication is expected to fall from its high of 80 percent to approximately 40 percent within the next decade.
>
> (Graddol, 1997: 61)

> Whereas more than 90 percent of the early users of the Internet were located in North America, the Net is now growing fastest in developing countries; in China and India alone, internet access is expected to multiply fifteen-fold over a two year period to reach 5.5 million users by 2002.
>
> (Warschauer, 2000: 156–157)

None the less, English remained the dominant language for programming conventions. English and romanized script is the glue of any global 'virtual community' creating a higher 'entry cost' for non-Anglophone users. However, given the much larger populations in countries such as China critics have argued that,

> in order to adapt technology to users, rather than adapt users to online monolingualism, a non-Anglophone internet must be developed. Reconceptualizing the Net as a set of linked human languages that actualize heteroglossic norms, not simply as a domain of market segmentation and software localization from an English norm, will aid in propelling minoritarianism from out of the electronic shadows.
>
> (Lockard, 2000: 177)

VIRTUAL AFRICA

There is a huge gap in the access to digital virtuality. The United States and other OECD countries have 15 per cent of the world's population but 88 per cent of Internet users. Globally, 80 per cent of the world's population has never placed a phone call. For those who do, in countries such as Ghana, a phone may be up to ten kilometres' walk. Africa has less than 2 per cent of the world's telephone lines: 70 phone lines per 100 Americans compared to 2.5 lines for every 100 Africans and less than one computer per hundred people (there are approximately 6 million computers on the continent, not necessarily all functioning). Estimates of the number of people in Africa who have used the Internet range between 4.15 million and 1.35 million.[1] 'More people use the Internet in London than in all of Africa' (figures from International Telecommunications Union, cited in Bray, 2001b: A24).[2]

The notion of a 'digital divide' is the latest expression of the economic chasm that separates the 'First' and 'Third' worlds. Indeed, this description has found new life on the basis of contrasting levels of development of information and communication technologies (Wresch, 1996; OECD, 2001). Huge investments by G-8 countries aimed at disseminating computer technology will likely be frustrated by the language barrier noted above. This is a far more difficult hurdle for most users. In most of Africa the hourly cost of accessing the Net is estimated to be US$14, compared with US$1.45 in the United States and under US$3.00 in most of the European Union countries (OECD, 2001).

> Pre-existing economic disadvantages translate into a new electronic differential, which in turn reinforces the old symbolic order through a new structural racism of limited or absent internet access. Because the internet additively re-enunciates the languages of power that have dominated the economic existence of entire nonwhite continents, it electronically amplifies and reproduces the modern history of capital.
>
> (Lockard, 2000: 179)

Opportunities emerge for those who are in a position to take initiative but those who hold power can extend their control and activities through the Internet. Yet the Net itself neither reduces food supply and security nor does it create poverty. Investments in communication infrastructure allow 'First' world firms to hire low-wage workers in countries which speak European languages. These are call centre operators and technical help line assistants (see previous chapters) in, for example, India. Fewer opportunities for high-skilled workers will be created. Attempts have been made for a decade to bring high-speed Internet and telephone to Africa. Satellites are expensive, limited in their capacity and switched through Europe. The history of unpredictable politics and climate make land-lines difficult to maintain,[3] and so an undersea cable has been proposed. However, this $US1.8 billion 'Africa One' project, or South Africa's SAT-3 cable servicing the West coast from the Cape to Gibraltar, is seen by many as folly: The 'system would be serving one of the least developed routes in the world', comments one critic (Michael Ruddy, Terabit Consulting, quoted in Bray, 2001a: A25). Still controlled by state monopolies, 'last mile' phone lines to businesses and homes are lacking, meaning that mobile phones are the preferred connection – but are available only in major population centres. The arrival of digital and mobile technologies create new competitors to old, state-run 'legacy' systems that often required hefty bribes and long waits for telephone and other connections. For example, take the case of Francis Quartey who returned to his native Ghana after working in the United States at AT&T:

> In 1998, he launched an internet service provider called IDN. In 2000, IDN began offering 'voice over IP,' the technical term for placing voice telephone calls over internet circuits. Because internet phone calls bypassed the state-run phone monopoly, Ghana Telecom, and its high rates for local and international calls, voice over IP could save some users hundreds of dollars a month. But it would also subtract that amount from the ledgers of Ghana Telecom.
>
> . . . One day they just walked in there with guns and stuff, and took me out, and the equipment, said Quartey. He spent three days in jail before a judge threw out the case against him. 'When I got my equipment, some had been destroyed, some was stolen.'
>
> (Bray, 2001b: A25)

The Internet has permitted the growth of a new African virtual class. According to UN data, this group is male, 25–35, university educated, English-speaking and associated with non-governmental organizations or the media. The well-known age and gender bias of information and communication technologies in favour of young men is amplified in developing-world conditions. Private dial-up kiosks or cybercafés, and free email addresses on Hotmail or Yahoo allow them 'to converse with one another across international boundaries and even within nations and cities with frequencies and in ways not previously possible . . . they share elite characteristics not previously discernible but identify with a heterogeneous diversity of organizations and groups' (Franda, 2002: 18; United Nations Economic Commission for Africa 2000 survey data). For those without connections in countries in Africa and elsewhere such as Sri Lanka, 'radio surfing' shows are broadcast on commercial radio (the dominant mass media in Africa). Listeners write in with requests for searches or topics and the radio hosts describe the results of Internet searches. However, the rise of a new 'elite' of users engaged with global flows of ideas and – in many cases – resources, while others merely listen in, suggests the exacerbation of existing inequalities.

Non-governmental organizations have been strengthened by the availability of the Internet, even if often only via small, privately-run market kiosks. Successful campaigns have been coordinated within countries and regions. It is also significant that global and mainstream Internet audiences in IRC chat channels and Usenet newsgroups have been tapped. Some see this as the beginning of the decline of state power (Matthews, 1997) but others point out that states can be equally enabled (Slaughter, 1997) as are collaborations sponsored by states and delivered by NGOs. Wu sees the Internet as just such a co-ordinated regime of interactions, not a disconnected cyberspace (Wu, 1997).

Flows of information also have political implications. *The Economist* reported that South African reluctance to allow the sale of AIDs treatments such as the drug AZT stemmed from the President of South Africa's discovery of anti-AZT information on the Internet. Reports of lawsuits in the US and the UK because of the drug's side-effects circulated widely on the Web and in chat rooms (*Economist*, 1999). This information was over-rated in relation to local reports of the rapidly-rising HIV infection-rate and death toll, leading to a significant delay in government intervention and a high cost of lives.

The critics of globalization fear that cultural standardization and the further penetration of Western popular culture world-wide will create new forms of dependency (but see Parker, 2001). However, others worry about the implications for the stability of repressive states as exiled elites use the Internet and new telephone connections to influence local voters. Isolated populations have been useful for many African politicians, but expatriates have discovered a sense of identity-in-diaspora via online cultural spaces such as chat rooms and newsgroups. These have been used to create a virtual sense of community (Silver, 2000: 136).

Opposition groups use the Internet to pressure the overseas African diaspora. For example, Ghanaians working abroad send over US$350 million back to families in Ghana. Wherever there are free elections, these foreign remittances can be used to cajole local residents to vote out governments. China and Saudi Arabia have made widely discussed attempts to ban cyber cafes and block websites. Users are viewed negatively as addicts indulging themselves. But a key worry for Chinese officials is the use of the Internet to post pro-democracy arguments on websites outside of China which are out of reach of the government.[4] A more profound problem, however, is that the Internet, with the possibility of users concealing their identities and encrypting messages, is a medium of distribution which is difficult to police.

> the diasporan web – nowhere more prevalent than in the free exchange of beats in musics of the African diaspora [and its hybrids] . . . has superseded the corporate model of hierarchical information flows. The discrete roles of sender, channel and receiver have been blended into a single complex.
>
> (Cubitt, 1998: 145)

However, the situation is still overwhelmingly negative. As Cubitt summarizes the scene at the end of the 1990s,

> The economic and statistical instruments that we have agree: in the Pacific Rim, the European Union and North American Free Trade Area [NAFTA], the rich (top 40 per cent) are getting richer and the poor (bottom 60 per cent) are getting poorer, . . . education, the virtual route to improving life chances in a changing world, is intensively geographically bound, so that depressed areas get the worst schooling, exacerbating the production of a permanent underclass. *Africa is supernumerary to*

> *the requirements of the global information economy*, and the vast majority of its population will be left to murderous factional struggles for control of the state in order to secure the scraps of aid that drip through it.
>
> (Cubitt, 1998: 133; italics added)

SUMMARY

This chapter sought to examine the relationship between economic globalization, diasporic communities of exiles and the manner in which Africa became an overlooked continent in the drive to create the necessary infrastructure for a global digital virtuality. This follows the existing logic of economic inequality. Gaps in access to digital virtuality may translate into new economic disadvantages, but flows of information also have political implications. Internet access adds a new register of inequality in developing world contexts which counters attempts to use the Internet to promote economic development and social justice. Censorship and persecution of postings on foreign websites are now common policies in totalitarian regimes. Exclusion from global networks perpetuates insular political states by restricting the flow of information from the outside, in particular from diasporic communities of exiles.

5

JOYSTICK GENERATION

Cyberpunks, camkids and family life

The digitally virtual reaches out beyond the realm of the Internet to impact on everyday lives grounded in the limits and frailty of the body and embedded in the experience of concrete events. This chapter considers the impact of digital virtuality in everyday life, focusing primarily on the home and recreational uses of digital virtualities.

- The blurring of the boundaries of the household and family time with the worlds of work, commerce and mass media.
- The competition for attention brought by an overload of information and the virtual enslavement of populations to communications technologies and flows.
- Fears and attempts to describe shifts in our relationship to the virtual as a cyberpunk subculture.
- Youth strategies for dealing with broader digital virtualities, and integrating them into leisure and social interaction – a joystick generation.
- The implications of virtuality as an everyday phenomenon.

The impact of the opening up of the virtual via digital technologies has been felt in the home. Although less obvious than the workplace in terms of machinery, the domestic sphere is penetrated by communications technologies such as the telephone and television and stuffed with toys and small devices in which chips are embedded, allowing digital forms of virtuality to spread quickly into children's and families' lives in North America and Europe. Even in countries which lagged behind the trend-setting economies, the take-up of the computer was one of the most rapid disseminations of new technology on record. Computers are available at friends' houses where they are used for entertainment, in school and in public libraries where they are used not only for information but for surfing – an interactive form of 'infotainment'. In the UK, for example, December 1995 was said to be the Christmas that parents, anxious that their children be included in the digital revolution, bought a home computer. Perhaps a winter of disappointment followed, as the supporting infrastructure was not in place to provide a user-friendly and instantly accessible virtual commons.

It was left to a generation of children and adolescents to figure out what to do with the equipment cluttering their desks, living rooms and the corners of suburban recreation rooms. Hollywood films such as *You've Got Mail* portrayed the implications of email and the anonymity of newsgroup postings and Internet chat for undoing and forming relationships. But it was another five years before computer workstations began to make their way into the kitchens and family rooms of the affluent West, where they had become indispensable aids to homework, to keeping in touch with relatives, making sense of and perhaps reprinting one's holiday photos, and playing games. The home computer and its software as a recollection machine, a database of memorable recipes, and as a simulator of virtual play environments has been integrated into domestic life. This may seem a middle-class perspective to some, but the darker side is hinted at in the anxiety over falling behind. Few have questioned the discourse of a 'digital divide' – this is not just a question of necessary skills.

Why must everyone become a consumer of computer technology, of online entertainment and of digital virtuality? Like it or not, computer skills are not essential to life. The reduction of family time, and the irrelevance of public–private boundaries of the home have not as yet been systematically examined.

CYBERSERFS IN CYBERIA

The virtual bursts into family life in the form of online videogames, Internet chat and concern over what materials children may be browsing and who they may be meeting online. Parenting advice books recommend that parents limit the time children are allowed to spend in front of a computer. In its most paranoid form, the Canadian government and missing persons agencies are sponsoring a hotline and 'tips' website (Cyber Wise) to recruit children and women to inform on 'husbands' and fathers who may be exchanging messages with 'very young girls' (Blackwell, 2002). However, other intrusions of the virtual into everyday family life are less well examined. Tempted by advertisers and by the self-interested generosity of managers, populations in North America, Europe and Asia have fetishized the latest gizmos. Pagers, cellphones, pocket email devices and telecommuting weave the virtual worlds of the Internet and other digital virtualities much more closely into family life than ever before. The public intrudes on the private. Even if they are 'interruptions', treated as conceptually and culturally separate from the domestic sphere, the intrusion of online fora, networked professional work environments and even the virtuality of distant callers on the phone (see Chapter 6) whips up the private space of the nuclear family until it foams with the semipublic bubbles of the commercial world and work time. These fora include messaging systems such as chat on IRC, instant messaging (IM) such as ICQ and more graphically rich online digital environments offered by 'massively multi-player computergames' (MMPGs – see below). Professional work environments include programs such as 'Lotus Notes' and also shared websites for group collaboration and file sharing such as 'Yahoo-groups', 'Groove.com' or 'communityzero.com' (see Chapters 6 and 7).

In North America, telemarketing is particularly common. Offers designed to hook people into listening to a sales pitch are made to those who answer computer-dialled phone calls. These are based on tracking purchases and 'data-mining' records of people's transactions and subscriptions, and even their use of utilities such as individual consumption. For those who do not answer, voicemail may be clogged with lengthy messages which are left automatically. Through such interruptions, the professional and market worlds of the public sphere have become more and more prominent in household life. In effect, telemarketing is the daughter of the virtual. 'Customer profiling' entails building up a model of typical users of services and the consumers of specific goods. This model, the ideal consumer of a company's product, is an abstraction which represents potential consumers. Not only existing clients, but the consumers of competitors' services and similar goods, are 'profiled' in databases of their characteristics – household income, age, gender, children, occupation, educational attainment. Each actual client therefore has a double; their profile is a virtual image. These *virtual customers* dwell in the databases of customer contacts and profiles, databases that are themselves virtual environments. Additional databases covering other populations are acquired. The objective is to assemble a census containing enough detail to stand in for the population as a *virtual market*. Based on their resemblance to the ideal, to what we might call an *abstract customer*, people are identified as potential clients and contacted. It is not unknown for telemarketing workers in the call-centres of large banks and other Fortune 500 corporations to call at 9 p.m. or later, or to phone, with uncanny timing, at dinner time.

Telework is a related issue (see Chapter 6). The use of devices and the right to work from home as a reward and as a sign of organizational status has led workers to acquiesce in becoming *cyberserfs* – *virtual slaves* to technology and to organizations because of their surrender of control over their own *attention*.

Work from home represents a struggle to renegotiate the work day, trading the 'opportunity' to monitor children or to work in a less stressful or more physically accessible environment for the intrusion of work-related calls which may come at any hour. In this struggle, the balance is slowly shifting from the worker's back to corporate managers. The job never ends. Or, it becomes doubly onerous as care for other family members (children, the elderly) is layered on top of pressure to perform, to concentrate on computer-based 'knowledge work' such as writing, correcting reports or inputting data. Workers acquiesce to become cyber-serfs whose situation resembles too closely the control which feudal landlords exercised over the private, family lives of their tenants. The peasantry was bound by more than economics. Social hierarchy and duty provided a disciplining ideology anchoring an unequal social order. Twenty-first-century cyber-serfs compulsively answer, respond and interact in part from a desire to feel needed and important, to be 'in the loop', to be socially central (Lefebvre, 1968; Maffesoli, 1996). Individuals sacrifice not only 'family time' but personal time – with un-documented implications for personal development and character in the future. What are the psychological, social and physiological implications for people without hobbies burning out at work? Some, such as GenX slackers, have attempted to reassert this balance but this remains a scorned and marginalized effort.

The abuse of patience, the demand for instant attention to queries from work and frustration with assignments passed down by email does not disappear once one hangs up or hits the off switch. Emotions carry over into family life – back to the dinner-table and to sleepless nights. Radical steps must be taken to escape; but virtual intruders count on the disciplining effect of politeness and manners. Interruptions during 'family time', weekends, requirements that one check in by email even during vacations makes them not only 'connected' but puts workers increasingly on call. Children receive an abject lesson in work-as-priority. The public and private spheres are no longer separated, as claimed by sociologists such as Jürgen Habermas for the case of 'modernity'. Sheller and Urry have suggested that this dualism no longer provides a useful basis for analysis of social issues. They point to the conduct of private business in public spaces on cellphones, the privacy in public afforded by the automobile, the privatized aural worlds of the walkman user in the streets. It is important to add, however, that this is an unequal mixing. The private loses to the public. Individuals and households lose the ability to control the conduct of professional business in the context of the private sphere, of family life and personal time (Sheller and Urry, 2000).

Cyberia, the popular term for this condition, is not a place (an abstract concept or a metaphor of suburbia) but a *predicament* – a virtual suburbia of our worst imaginings, populated by cyberserfs young and old. Cyberia as a popular term marks the extent to which people understand that they live in a different world from the twentieth-century suburban ideal. Enshrined in media images, suburbia is a museum piece looked upon with nostalgia by those who know that their roofs offer little shelter from the rain of digital information. In effect, suburbia has become a topic for architectural and urban historians. The aluminium-clad bungalows of 1970s North America look like such simplified social environments that their calm is enticing. Today, like out-of-control eighteen-wheelers speeding down residential streets, commercial websites offer children free games in return for youth consumption information. Children playing digital forms of Pokemon, however, learn that these are rough streets. Like an unmonitored playground, the older children send threatening email, and hack the 'accounts' and collections of the younger ones to steal their characters and alter their avatars (*Citizen*, 2001).

In this porous and mixed material/virtual world of the home, parents worry that children and teens will be targets for online predators, paedophiles and online marketing scams at home. Columnists and how-to authors warn parents to limit their children's online time, double-check their email for pornography and debate whether or not face-to-face meetings with others met on IRC should be allowed (Aftab, 2001). Digital virtuality expands the directions from which encounters may come, whether friend or foe. This has brought anxiety to many parents, who find themselves ill-equipped to evaluate online contacts and concerned about the opening of family life to influences from new spaces. Digital virtuality has its rough streets. Like a cyberspace version of the tale of Romeo and Juliet, friends and influences may come not only from social groups that are disapproved of, perhaps 'the wrong side of town', but from the other side of huge cities and states.

INTERNET PROOFING

Addicted to interacting through computers, a generation of youth has been called 'Otaku', the derogatory label proudly adopted by devotees of Japanese 'anime' cartoons, cyberkids and cyberpunks (Beineix, 1999). Like 'street proofing' (instructing children on precautions to be taken when dealing with or meeting strangers), computer-mediated communication in the home brings with it a need for 'Internet proofing'. The difficulty is that parents, as busy late adopters, are in a poor position to guide children, who are adept early adopters of the technology and media, but often have no experience or preparation for encounters with adult and older children's understanding of things as simple as the irreversibility of the sale of a comic book or collector's card, or any sense of the manner in which value is set by a market larger than a buyer and seller. However, 'Instant messaging, which has become a major concern of many parents, has been a blessing for some teens and their families – particularly among the hearing-impaired' (Thomas, 2001).

Instant messaging (IM) puts old TTY text-phone technology for the deaf to shame. Online communication via chat in real time and pop-up instant messenging such as AOL's AIM (which interrupts another user to whom you wish to send a text message or image) has exceeded expectations in facilitating social interaction for the hearing-impaired and others who have found themselves on an unequal footing in the embodied social worlds of the school or neighbourhood.

The 'compulsion to proximity' operates because people seek to broaden their contact with others. For example, online dating services often stumble on the difference between virtual and material meetings. What sounds good on email fizzles in 'real life' – and in fact does not even make it beyond the first meeting into everyday life.

> 'What we find is that people like good, old-fashioned meetings in person . . . I don't think computers will ever replace that' says Nancy Slotnick, owner of Drip Café in New York and her Lovelife Dating Service. The service at www.dripcafe.com offers not only online profiles of potential dates to monthly subscribers but face-to-face meetings in a network of 10 cafes in Boston and New York.
>
> (quoted in Wright, 2001)

The fear of meeting undesirables may attest to the paranoia and control manias of boomer parents faced with new lines of escape (see Witmer, 1996). Twenty-first-century youth has at its fingertips new badges of identity to distinguish themselves from others – whether their parents or the moronic-seeming countercultures of the past, strung-out on hash and coke. The digitally virtual in cyberspace offers all-absorbing escapism of role-playing and battle games such as *Myst* and *Doom*, respectively, plus new ways of congregating out of parental earshot and contact with forbidden others away from family and societal surveillance. Although meagre in their bandwidth and impoverished compared to embodied interaction, IRC chat, newsgroup postings and web homepages offered an alternative social sphere to many people's experiences of marginalization at school. A social teen phobia had sought to exclude youth from commercial and public spaces whether malls, parking lots of fast-food outlets and convenience stores, or public places. Police forces, municipal governments, schools and parents scrambled to close up the counterspaces which sprang up in cyberspace with a devotion matched only to attempts to render material public spaces unfriendly to skateboarders.

These are zones outside of the equation of both public and private duties. As such they offer the opportunity for escape and the possibility of experimenting with alternative social norms. In this sense they are what may be called *counterspaces* – 'alternative spatial systems, arrangements, practices, norms at work around us, with all their ambiguities and failures' (Lefebvre, 1981: 443) – where the tendency of social space to escape bureaucratic management was exploited. Counterspaces are social creations which build on the *liminality*, or 'in-betweenness' of digital virtuality and its tendency to escape from the imposition of social norms (see previous chapters; see also Shields, 1991, ch. 2). Bey dubs these Temporary Autonomous Zones (TAZs) pirate utopias (Bey, 1985).

Headlines such as 'online Lolitas flirt with anonymous voyeurs' mark the media's moral panic over how web cams could be put to use by young women. They are now widely available and often sold 'bundled' with other products: 'Flash a bit of tummy, a little cleavage, or more, and you'll hook yourself an online sugar daddy: That's what a host of teenagers are doing as they go "whoring for hits" – and gifts' (Mieszkowski, 2001: B1). Yet the evidence is still scant. The images are for the most part time-delay snapshots of life in the adolescent bedroom, studying, reading, sleeping, shot from miniature television cameras perched atop computer screens. These are 'urban myths' about 'a friend of a friend', always someone else who has raked in gifts from online fans who egg on these underage 'cam girls' to bare a bit more. Online gift lists and wedding registries work as ideal 'wish lists' for cam girls because merchants generally don't reveal the 'wisher's' location: 'the relationship between the online pen pal or fan can remain entirely virtual, yet still produce the goods' (Mieszkowski, 2001: B1). The images are also supplemented by daily diaries and typed, two-way chat, a feature pioneered on the early web cam servers such as CUSeeMe. The cam sites are perverse in their eradication of privacy and 'true-confessions' style, and 'obscene' not in their content but in Baudrillard's sense of over-exposed, or overly visualized: *ob* (over)-*seen*. The over-exposure of the private sphere is redolent with adolescent discontent over the limits of their world and the ability to exert control over scripted and dominated lives.

This is a radically different approach to the personal and private from the secrecy and encryption required to maintain the privilege of privacy and the private sphere as a limited 'temporary autonomous zone' of individual freedom – in the bedroom, for example. Instead of the invisibility of the private, those who live their lives on camera or those who selectively expose intimate moments (even if they are staged, posed and simulated intimacies) practise a tactical super-visibility. Edited down to the visual image of the web cam, the virtual becomes more banal than the actuality and material weight of everyday life. Amidst the over-exposure and clutter of detail, the single crucial detail – the secret – may be easy to miss. As Abdou Malik Simone commented in a presentation on the public life of African markets, how does one move contraband through the open stalls and the inquisitive and competitive surveillance of others in the same business? Right under their noses, in open daylight in the most nonchalant fashion (2001). Over-exposure as a form of banal 'ob-scenity' (or super-visibility) hopes to slip through the analytic capacities of human and automated systems which may be overwhelmed by the minutiae of everyday life.[1]

The web cam girls and boys are young, often appearing to be under the age of 18. Their fantasy is possibly one of a 'Santa Claus [who] is going to come along and take care of you and not expect anything' in return. 'The cam sites represent the kind of risk-taking teens engage in to form their identity' (cited in Mieszkowski, 2001). One peer, Bridget Therese Guildner, posts the comment,

> I think that these girls are just now discovering that they can make men do things . . . and especially say things just because

> the men think that they are desirable. . . . However, I think that the reason why we don't see many 25-year olds running cam-girl sites is that with experience comes the realization that being used is unpleasant, even if you are using the person back.
>
> (cited in Mieszkowski, 2001: B2)

Setting aside moralizing reactions, web cam sites are significant as diaries and statements of desire – wish lists of commodities rather than life goals. Cynically, we might say that these substitute for a life defined in terms of social meaning while the camera broadens the bandwidth of online social interaction, but allows just a one-way visual presentation of self so that one does not have to look at the fans who remain imagined, virtual. The images of cam girls are more cleavage than breasts; more study-hall portraits than peep-show nudes. Any view is merely an insight into the boredom of the bedroom computer addict; close-ups of interminable, fatiguing hours in front of a screen, keyboard, mouse. 'Some of the young girls really aren't showing that much skin' (Marissa, cited in Mieszkowski, 2001: B2), but in the era of AIDS and fear of the other, one can imagine or 'project' one's other while offering an image, possibly a titillating one, of oneself. A stranger or another independent individual is unpredictable, uncontrolled. Confining such otherness to the virtual becomes a strategy by which the terms of sociality may be set and controlled by youth.

At school, research shows that the location of school work shifts decisively from the school to home where many families now have computers. Schoolchildren perceive computer-formatted projects to be more highly rewarded (despite teachers' protests to the contrary), and work with multimedia sources (such as Microsoft's 'Encarta' encyclopaedia on a CD-Rom) is more fun. New technology changes people's practices, their ways of understanding the world through representations (such as a virtual 'globe' in 'Encarta' or human anatomy in animated software such as 'The Virtual Body') and hence the way in which they go on to create the world (Winograd and Flores, 1988).

School work itself shifts to a form of bricolage: students approximate, 'they tinker, tweak, learn from their mistakes' (Cuthell, 2002: 132), and they do this on their own by watching other children perform specific tasks, leaving projects incomplete and working on other things until they have learned what the 'trick' of a program is.

> They go for Best Fit in order to make applications work together, data integrate and meet their deadlines. The learning patterns they are developing are not those of their teachers, but they work.
>
> The more they use these strategies . . . the more they are likely to apply them to situations in which their teachers have predicated the learning on more conventional patterns . . . many students felt that the site of learning . . . was at home, in front of 'their' computer.
>
> This is the fundamental problem, which must be addressed by both teachers and the educational system. If school is not seen as the principal site of learning work for the majority of students, then what is it for?
>
> (Cuthell, 2002: 132)

Cyberkids are autonomous learners who seek approval from their peers, mentoring and advocacy from teachers (Schostak, 1988), and clear performance criteria that are not dazzled by IT skills but can specify the goals of learning.

CYBERPUNKS OR VIRTUAL SUBCULTURES?

In William Gibson's fictional *Neuromancer* (Gibson, 1984), cyberspace has its own forms of beauty and of crime – rebels and outsiders whose data trespasses are deterred and punished by lethal shocks to their mind-to-computer neural interfaces. In this context, *cyberpunks* were identified with the outsider-heros of these novels and Japanese *anime* comic characters who fought evil techno-villains (Rucker, 1992: 64). Glamorized by magazines such as *Mondo 2000*, *Wired*, and hyped in cover stories (*Time*, 8 February 1993), analysts saw a youth subculture aspect to the mundane activity of surfing the Web. However, this was blurred with, and overshadowed by, the hacking community and video-gamers. MUDS and other interactive textual environments (early forms of shared virtual environments) were integral to all these groups. They were, and are, used for public discussion and role-playing games and create a sense of online, or virtual, community. However, the significance of hackers and cyber-punks is that a sense of identity and community depends on an affiliation with digital technology and fetishized programming skills.

> For the first generation to grow up with computers in their homes, technological access to electronic information networks is a natural condition of the domestic scene. Having claimed cyberspace as their own private frontier, cyberpunks resent the imposition of limits on their cyberspace travels . . . on the one hand, cyberpunk subculture popularizes a fantasy of resistance and opposition to corporate information control, it also projects a fantasy world where the material body – the race, gender and ability-marked body – is technologically repressed.
>
> (Balsamo, 1995: 348)

Youth subcultures had perfected the twentieth-century 'escape attempt' around consumer identities woven out of two elements – identifying objects and behaviour (see Shields, 1993). Consumption objects such as particular forms of music, clothing, drugs and motorcycles were taken up by rebellious teens and consolidated by marketers tracking youth tastes. One old nugget, the early to mid-1960s British 'Mods', became a goldmine for sociologists and a key pop culture reference. 'Mods' were pegged as the upwardly mobile children of blue-collar manufacturing and service workers, who attempted to differentiate themselves from their working-class parents through style. They favoured modern Italian fashion and streamlined scooters rather than the French and English mopeds and motorcycles whose mechanical image underlined the contrast with the more traditional values of rival subcultures of 'Rockers'. Iconic behaviour such as weekend scooter rides to the British south coast became rites of passage, as did participation in seaside riots made famous in the ensuing media panic (see Shields, 1991). The Mods provided a template for later subculture's use of style and taste as badges of identity, such as the Punks of the 1980s.

By contrast, the late 1990s witnessed the virtualization of subculture as identifying badges, and consumption patterns and behaviours became more symbolic, such as listening to music or hacking establishment systems, and went online, via chat. Were the new rates of Internet use, online chat and gaming part of a new subculture? In North America, some attempted to pigeonhole youth engagement with the digital virtuality as 'cyberpunk', but this flat-footed effort misconstrued the virtual and focused on the less significant material markers typical of the older twentieth-century youth subcultures – clothing, decor, location (staying at home in one's bedroom) and machinery. But the argument here is that the virtual is of global cultural significance, not a subcultural matter, even if youth and the boosters of cyberspace were the first to grasp its potential. Authors concur that the defining quality of cyberpunk is/was an irreverent attitude towards limited access to stored data as private commercial property (Balsamo, 1995). 'Information wants to be free' was the slogan of the time (cf. Hughes, 1993).

However, the true defining quality of the cyberpunk, including hackers, crackers and various other self-styled 'freaks', was and continues to be digital virtuality as a leisure-free space. Nowhere is this more obvious than in the case of video-gamers. This has long been a sociable activity, not just a solitary game played like computer chess. The first computer game, *Spacewar*, was written by an MIT student for not one but two people and played on the university's recently acquired Digital PDP-1 (see Chapter 3). 'Duelling spaceships shot at each other' with releasing blobs of light as 'photon torpedoes. Four buttons linked to a video display terminal controlled each ship.' The code was freely distributed and was soon installed on almost all the research computers in the early ARPA-Internet, the predecessor of the Internet. 'Nearly every coder in the country played this primitive form of electronic dodge ball. In an age when computer time was a valuable commodity, Spacewar would be responsible for millions of dollars worth of diverted processing power' (Shulman, 2001).

VIRTUAL FEARS: HACKERS AND REAL KNOWLEDGE

The Internet has come to be the locus for the expression of the sense of insecurity in everyday life. In one sense, fears over online theft and misuse of personal passwords, credit card and bank account numbers reflect all the fears of contemporary consumers. This includes heightened problems of credit card fraud and the theft of bank card personal identification numbers and other access codes. It is not that these fears are misplaced but that encrypted online transactions are no less secure than other transactions. Fears over Internet security are iconic of the situation of life in risk society.[2] A moral panic over hackers, whether 'good' or 'bad', spills over into vilification of those who offer online encryption (Hum, 2001). Hackers are feared to be gaining unimpeded access to corporate and personal information through wireless networks set up in temporary offices on which security features have yet to be activated. Security consultants and manufacturers make comments such as, 'The problem is that you can have people access some of these networks with an antenna and a Pringles can' (Nick Tidd, 3Com Canada, quoted in Thompson, 2002), evoking a now stereotypical image of the Net as a space of risk (see Chapter 8), and of error, untruth and uncertain threats.

Most analysts worry that people's resort to the virtual introduces psychological changes in which virtual data and environments are treated as 'real' – by which they generally intend 'actual'. The virtuality of the Web threatens 'the real' for conservative commentators. Others worry about the circulation of inaccurate information, that

> The Net is a medium not for propaganda but for conspiracy . . . which allows all kinds of people to enter the conversation . . . as people move onto the Net, they tend to lose their common sense and believe all kinds of crazy tales and theories. . . . We take a story's appearance online, as well as in print, as proof that it has been subjected to rigorous journalistic standards, but there's so much stuff out there that no one has the time to contradict all the errors.
>
> (Dyson, 1998)

Jodi Dean incisively diagnoses an anxiety over the inclusive public sphere in these comments: 'Who exactly loses her common sense . . . ? Presumably the ignorant, ill-informed . . . those left unguided by . . . entrenched authorities' (Dean, 2000). Those who aren't 'us' – and Dyson's us is a select, professional group only. But 'who today shares her confidence in journalistic standards' (Dean, 1998: 65)? Dyson's comment suggests the prevalence of an ancient prejudice against the oral which is dismissed as hearsay and rumour. 'Real knowledge' must be legitimated, official information (and hence concrete); knowledge always involves a process of reflection and knowing, which involves the virtual and abstract (see Chapter 1). The more concrete the better: published writing is the ultimate reliable source of 'truth'. In later chapters we will return to the narrow conception of trust that this notion of truth implies (see Chapter 8).

> In her anxiety around the inclusion and access the Web provides, Dyson returns to an eighteenth-century conception of truth and concomitantly narrow assumption of trust. She posits a field of knowledge deemed reliable precisely because of the credibility – to her – of a small group of authorized, trusted, speakers. Only a few can be believed, only a few produce 'real knowledge'.
>
> (Dean, 2000: 66)

JOYSTICK GENERATION: VIRTUAL SUBCULTURE

Writing with tongue in cheek in 1982, Martin Amis recounts his first encounter with the *Space Invaders* video-game in a bar:

> Grunting heavies were wrestling with what looked like a sheeted refrigerator. They installed it in the corner, plugged it in, and drew back the veil. The invasion of the Space Invaders had begun.
>
> Now I had played quite a few bar machines in my time. . . . But I know instantly that this was something different, something special. Cinematic melodrama blazing on the screen, infinite firing capacity, the beautiful responsiveness of the defending turret, the sting and pow of the missiles, the background pulse of the quickening heartbeat, the inexorable descent of the bomb-dumping monsters: my awesome takes, to save Earth from destruction!
>
> Now after nearly three years, the passion has not cooled. I don't see much of Space Invaders any more, it's true. . . . These days I fool around with a whole harem of newer, brasher machines. When I get bored . . . a younger replacement is always available. . . . The only trouble is, they take up all my time and my money. And I can't seem to find any girlfriends.
>
> (Amis, 1982)

Parents fear the more recent games such as *Quake* most, because they marry the role-playing of Multi-User Dungeons (MUDs) with the graphical sophistication of video-games designed to run on dedicated platforms such as Sony's PlayStation. These games allow thousands of players to exist in the same virtual environment. Players can meet each other, adopt the roles of various avatars and build their character by gaining points and entitlements from experience in the game. They can also 'cooperate with others to accomplish goals or just hang out and talk via typed messages to the other players, who invariably are scattered across the world', writes one local commentator (Shulman, 2001) (see Table 5.1 for a history of video-games). These are the games that almost demand hours of concentration on exchanges via telephone and ethernet. Game. Eat. Game. Sleep. Game. . . .

Table 5.1 Dates in the history of video-games (names trademarked, respective owners)

1962	*Spacewar!* (Steve Russell, Shag Graetz, Alan Kotok) – written by students at MIT for the Digital PDP-1
1971	*Computer Space* (Nutting) – an arcade game console with a 13" black-and-white-screen; proved too difficult for patrons.
1972	*Pong* (Nutting for Atari) – for arcade machines;
1976	First video-game consoles with screens for home use make their appearance (Vectrix).
1978	*Space Invaders* (Midway) – compelling game of destroying rows of 'invaders' descending at an ever-increasing pace.
1979	*Asteroids* (Atari) – all-time best-seller, introduced the high-score list displaying initials.
1981	*New England Journal of Medicine* reports on 'Space Invader wrist' repetitive strain injury.
1982	Colecovision platform – makes first use of the television set as display. *Donkey Kong* and *Super Mario* (Nintendo for the Colecovision).
1985	*Tetrix* (Alex Pajitnov) – written for PCs as well as arcade consoles.
1989	*GameBoy* (Nintendo) – portable version of popular Nintendo games. Sega Genesis platform expands home market.
1992	Concern over *Mortal Kombat* (Midway) – US Senate holds hearings on violence in video-games in 1993.
1993	*Doom* (id Software) – introduces many-to-many online environment or MMPG (massively multi-player game) mode. Sells 30 million copies (compared with 27 million official copies of Windows 3.1).
1993	*Myst* (Cyan and Cyan World) – best-selling computer game based on virtual adventure and role-playing.
1999	Early home video-game machines and games such as Atari had become widely sought after by collectors.
2000	PlayStation platform (Sony) – introduced with improved speed and graphic quality.
2001	*Final Fantasy* (Squaresoft) – the first digitally animated cross-over movie under the same name, after attempts at having humans act the part of video-game characters (e.g. Angelina Jolie as *Tomb Raider*'s character Lara Croft). Xbox platform (Microsoft) – introduced with integrated modem for participating in MMPGs over the Internet.

A spoof of a new computer game published in *The Net* magazine (2000) dubbed 'The Adventures of Fakk2' lampoons the shoot-'em-up character Lara Croft of Eidos' *Tomb Raider* game for PC, Nintendo and Sony PlayStation and the overblown colour and artwork of game packaging intended to appeal to young men: 'Play Dress-up'; '20+ Mind-Blowing Weapons – Battle Axe, Flamethrower, Rocket Launcher, Chaingun – you'll never run out of ways to kill.'

This online gaming magazine advert is actually a placement featuring the black humour and parody of *Heavy Metal* comic magazine. The violence and gore of video-games has provoked concern. In part this content developed as the first video-game generation matured, and demanded not only more challenging but more realistic and shocking material. As one video-game reviewer put it, it was Sega's Nintendo and later the Dreamcast that made video-gaming appealing 'to the rebellious and the angst ridden, the teens and the college crowd who wanted raunchy and raucous interactive entertainment, not leaping plumbers and plooping mushrooms' (Conlin, 2001).

Not only is the content of the Internet feared, but the virtuality of the medium and the toll that its use may take on the body is the focus of much worry. Children's lack of physical activity and their 'addiction' to playing video-games extends the preoccupation with fears of the impact of excessive TV watching and the resort to children's videos as 'electronic babysitters'. Although children rarely sit still when watching television but mimic dances and action that they see on screen, no matter how carefully crafted and vetted for psychological stimulation are television series and videos such as *Barney* and *Teletubbies*, there are still fears that the shifting content which appears to many to be 'play' rather than rational education (such as the counting games of the long-running *Sesame Street* series) or developmental exercise will dull children's perceptual sensitivity and intellectual sensibility. Virtuality of the on-screen image is intuitively contrasted with the materiality of action.

However, this too has been challenged. Computer systems such as *The KidsRoom* with motion sensors, sound and light equipment to project images on to the walls of a room may allow action to be moved into the material space of, for example, a child's bedroom while allowing several people to engage in an interactive experience which combines both real and virtual objects (Bobick, 1999). Operationalizing the virtual represents the latest stage in this subculture. Just as the film *Matrix* reverses the thesis of the virtual by positing that the actual is a virtual hoax in which digital technology may allow some to find flaws (discussed in Chapter 1), a new genre of video-game reflects back the paranoia of dominant culture over the virtual. These are intended to augment the virtual by blurring the lines of the actually real and the abstract in a manner resembling paranoia. This is done by using everyday forms of communication (telephone, fax) to augment the online, multi-player and interactive elements of the games. A phone call made by a computer-synthesized voice gives necessary clues or warnings once the game producer's computer detects that a player has reached a given point in the game (Shulman, 2001).

The fear of the video-game harkens back to Victorian commentators' fears over women who became absorbed in serialized novels by then notorious writers such as Dickens. If the generation of the 1920s and 1930s was the first to be affected by movies, the post-Second World War generation, such as the Mods, were the first to gravitate to the vinyl LP and to Rock 'n' Roll on 45s. The television first defined the baby-boomers, then it is the video-game and Internet that define twenty-first-century youth subcultures – not only gamers, but young music consumers using Napster, and Gnutella to download music and movies.

In his *Effect of Videogames on Children*, Gunter surveyed the field in 1998, concluding that playing video-games could boost problem-solving ability, whether or not those problems involved computers. The trial-and-error way in which many play video-games was correlated with scientific processes of exploration. 'In a world increasingly dominated by computers and requiring of people generally a higher degree of computer literacy, video games can, through the cognitive skills they demand, serve to play an important part in children's . . . intellectual and social development' (Gunter, 1998). Herz writes:

> Those to the joystick born have a built-in advantage. . . . Kids weaned on video games are not attention-deficient, morally stunted, illiterate little zombies who massacre people en masse after playing too much *Mortal Kombat*. They're simply acclimated to a world that increasingly resembles some kind of arcade experience.
>
> (Herz, 1997)

Like the music industry of the 1960s, the computer-gaming industry is on a break-out course. It aims to be the dominant entertainment medium of the twenty-first century. Video-gaming now rivals the movie industry in worldwide sales, with revenues around US$10 billion. In Canada, 21 per cent of the country aged 12 and over play computer games. If games such as computerized chess and Microsoft Window's Solitaire are taken into account, the average age was 28 in 2001. An estimated 179,000 between the ages of 25 and 34 play heavily six times or more per week. 'The most popular videogame of 2000 reflects that mature audience: Electronic Art's *The Sims* involves managing a household rather than ripping out the throats of slobbering ghouls' (Shulman, 2001).

Yet for children there is a physical and physiological impact to the sedentary life of the video-gamer. Rates of obesity have risen and physical fitness fallen. The overplaying of video-games leads to various repetitive strain injuries nicknamed 'joystick digit' and 'mouse elbow' (Osterman *et al.*, 1987; Mirman and Bonian, 1992; and see following chapters). Those who use vibrating 'force-feedback' joysticks may experience the hand–arm vibration traumas found usually only among those using high-powered vibrating tools such as jackhammers. Although manufacturers suggest limits, specialists worry that many play games for extreme lengths of time.

The joystick generation follows the media-created 'GenX'. But the new technology contributes an intergenerational schism even if one is sceptical of the breathless claims of Jon Katz writing online at slashdot.com: 'What's evolved is perhaps the widest gap – informational, cultural and factual – between the young and the old in human history. In many ways, gaming is at the centre of this chasm' (Katz, 2001). Popular ignorance separates non-video-gamers from devotees of the joystick. The culture of video-games is argued to be more profoundly different than a new genre of music or type of haircut. According to pundits, it spawns its own media culture while crowding out other media such as traditional book publishing. Video-games are now an integral part of the various toys owned by North American children, but are overwhelmingly identified as a boy's toy. However, although video-games have become big business and a major form of entertainment, barriers remain to granting them the cultural respect of films and pop music (both very much assembled by teams and edited or mixed from multiple 'takes').

EVERYDAY VIRTUALITY

In everyday life, 'we count on things to keep their place' because 'we have lived our lives with them in this fashion . . . in their fidelity to us they function as extensions of ourselves' (Romanyshyn, 1989: 193, quoted in Hillis, 1999: xxx). Things remaining in place help root our sanity. Just as the virtual becomes more a part of everyday thought processes, so everyday life is mixed up in the digitally virtual. Although serious educators and writers may lament the trivialization of online environments – the virtual dating, adolescent fantasies, relationship advice and virtual reality golf simulators are all signs that VR is becoming mundane.

This is not some sort of imported materiality but rather a deep need and willingness to rely on the *effectiveness* and virtues of the virtual. What was dubbed the 'postmodern condition' from the mid-1980s on is a recognition of the rapid pace of cultural and economic change. The *metaxis* of the 'virtually-so' and the essentially 'there' allows us paradoxically to seek stability in the digitally virtual when it is missing in our everyday lives (see Chapter 2). But this is likely to only 'heighten insecurity, having the effect of tilting cultural inclinations ever more strongly toward technical fixes' such as the digitally virtual (Simpson, 1995: 140; see also Chapter 8).

SUMMARY

This chapter has considered the impact of the virtual in the form of computer games and customer profiling in the context of the struggle to maintain control over family life and the home as a private domain. Videogames as entertainment, Email and chat as interpersonal communication are the most ubiquitous forms of the virtual that enter into the everyday life of families. Struggles for the control of attention and time appear pronounced in problems of family life and parenting. This appears as new forms of fear and risk, exemplified by the moral panic surrounding cyberpunks, and online pornography. Early attempts to describe online communication highlight the fear of contamination of the domestic and of morally policed public spaces by foreign and unknown elements.

Those who became involved with online media and with video-games in particular were at first described in terms of subcultures. However, the spread of video-gaming technology and the importance of digital virtuality rendered cyberpunk in to a key and growing element of contemporary Western culture. Rather than a true subculture, cyberpunks heralded culture-wide changes in the realtionship to the virtual.

6

WORK

Virtual working

In earlier chapters I have argued that virtual reality and even basic office work with the Internet relies on comfort with the virtual and through digital technologies creates a digital-virtual space. Much attention has been devoted to the small group of dot.com millionaires, the companies, and to the legal and administrative functionaries and decision-making structures that allow the Internet to function across diverse types of equipment, operating systems and languages. This chapter focuses on the technical and support workers, and the skilled labourers who work with the virtual on a daily basis. This chapter traces the development of this process and the manner in which the virtual is dependent not only on technology but on the human labour of technicians or 'Internetworkers' who stitch together the various technologies and institutionalized communication networks which allow virtual environments to function as simulations and as media or spaces where communications and data may be exchanged (Downey 2001). In outline, we will consider:

- The rise of virtual forms of work, and issues of status, alienation and information overload.
- The changing role of technicians' and support workers' careers to digital and other virtualities.
- Changing skills in the trades and professions.
- The experience of work at a distance with digital virtualities and in virtual teams.
- The impact of computing in the workplace, notably on clerical and secretarial work.
- The physiological impact of computer use on workers.
- The growth of teleworking from home and on the road as an example of virtual work.

A number of writers have pointed out the effects of the computerization of the workplace and the impact of virtual working. New forms of information, such as direct access to a database of sales which allows managers to understand better the performance of sales people, calls for and leads to new changes in the practices and routines of organizations (Nonaka and Takeuchi, 1995). The virtualization of work is a term used to cover a number of themes. First among these is a concentration on information and the use of information technologies to manipulate products which are tailored to clients based on information supplied or are primarily symbolic and informational to begin with (advertisements, educational and training materials, software, services such as accounting and payroll administration and so on).

> The very business itself is information. Many of the employees in any corporation are involved in the process of gathering, generating or transforming information.
>
> (Davidow and Malone, 1992: 65)

Virtual working involves more than information processing and what has been misleadingly dubbed 'knowledge work' rather than work with and on information (Zuboff, 1988). These two shifts drive a tendency to reorder and redistribute work within an organization, and to subcontracted specialists and computer experts (Nandhakumar, 1999). Just as the legal boundary of the firm is broken down by this contracting-out process (McLoughlin and Clark, 1994; Harris, 1998), so too are divisions and hierarchies within companies. During the last two decades of the twentieth century, there was a concerted effort to eliminate the mediating, organizing 'middle management' level of companies, with coordination rendered by the routine requirements of computers and the structure provided by requiring all aspects of work to be entered into databases. These acted as a form of surveillance over the progress of tasks (Elmer, 2002). In large organizations, geographically separated branches could work together on projects via video-conferencing (first via phone, then via the Web). Work which had always been done in a range of settings and from a variety of locations (work taken home, done in motels, aeroplanes and so on) could be formalized as 'telework'. It could take place in 'real time' as an active engagement with current projects, files and information on home computers networked over telephone lines. It could also be 'time-shifted' to take place around the clock in successive time-zones around the world to achieve more rapid progress, or to provide around-the-clock technical help. Hence the 'virtual office' and virtual organization. Employees also sought advantages in gaining control over the scheduling that 'flexible work' allowed, although this was rapidly erased through less progressive management, the application of piece-work pay contracts, and computerized monitoring and pacing of the work. Beyond workforce flexibility, more responsive, agile organization (Hale and Whitlam, 1997: 3) was sought via a shift in emphasis from structure to the essential processes associated with creating an organization's services or products (e.g. business process re-engineering; see also Tapscott, 1995; Grover and Kettinger, 1997).

Information technologies and the creation of digital virtualities may first be used to substitute for old technologies in existing ways of working, but the more significant implication has been seen to lie in the blending of expertise and enterprises across distance and time-zones (Jackson and van der Wielen, 1999). The strategy of the time–space network of less fixed, project-based 'virtual teams' (Lipnack and Stamps, 1997) contrasts with the ambition of many twentieth-century managers to create centralized, visible organizations, based in 'head offices' with a nine to five corps of workers employing their own self-contained expertise.

> Organizations used to be places. They used to be things. . . . But, as information technology catapults us into the reality of an Einsteinian world where old structures and forms of organization dissolve and at times become almost invisible, the old approach no longer works. Through the use of telephone, fax, electronic mail, computers, video, and other information technology, people and their organizations are becoming disembodied.
>
> (Morgan, 1993: 5)

This shift to more loosely organized work groups across companies requires a new attention to issues of trust, the negotiation of uncertainty, team culture and the latent aspects of organizations. Beneath the metaphors of disembodiment and the hype which proclaims the end of hierarchy is an attempt to beat the constraints of time and space (Mirchandani, 1999). In more theoretical terms, there are many similarities between digital-virtual workspaces and other 'imagined communities' (Anderson, 1983). However, shared, online and other virtual workspaces must still be grounded in the world of institutions, laws and cultural expectations. They must depend on infrastructure including wiring and buildings, and must cope with the challenges of time-zones and cross-cultural and interorganizational tensions.

This chapter considers the contradictions of workers faced with the simultaneous virtualization of work, reorganization of the material place and context of their work. Here the focus will be on the impact of digital virtualities, such as shared online workspaces with chat and filing capabilities, on virtual meeting rooms and the rise of teleworking from home and, more significantly, from cars, satellite offices and hotel rooms. Research tends to treat these as pure, spaces of technology and as spaces of information flows, rather than as what they are: unequal, *virtual spaces of labour*. They are stratified by power and status, privilege and varying competencies. Different workers play different roles. Lower-status technicians and support personnel in geographically remote 'back offices' or call centres maintain the illusion of seamless networks and flows of information between skilled users and machines and between professionals.

Most studies of the Internet do not consider workers at all. Some groups of workers are crucial to the implementation of digital virtuality, while at the same time they are impacted by others' decisions and actions. Only by examining their role and the limits to their agency can a deeper understanding of the changing world of work be attained. Few if any are 'only' users.

VIRTUALIZED WORK

The workplace is being 'virtualized' in ways that are both subtle and direct (Table 6.1) (Grover and Kettinger, 1997). An exchange that may once have taken place face-to-face or within the walls of an organization may now take place electronically via email, web-based chat or video and even automatically, as suppliers' databases display the use of their products by a client and update the required daily production-run. Jackson argues that these changes are driven by:

> The need for organisations to improve innovation and learning will demand new knowledge management systems, making use of IT support, that help members to acquire, accumulate, exchange and exploit organisational knowledge. . . . Because access to and transfer of knowledge and expertise will

> increasingly take place across boundaries (both organisational and spatial), internal networks and dispersed project groups, as well as inter-firm collaborations, will become more and more common.
>
> (Jackson, 1999: 2)

Often, tasks are now split between the materiality of physical labour – for example, restocking store shelves – and the virtuality of stock-taking, updating inventory lists and ordering

Table 6.1 Timeline: Virtualization of office work

1775	Card index (Paris Academie des Sciences)
1850s	Systematic library catalogues (British Library and others)
1872	Wooten roll-top desk with pigeon-holes patented
1920s	Vertical filing cabinets and folder systems for imperial paper sizes
1946	ENIAC 1
1951	UNIVAC computer (First commercial computer)
1954	Magnetic drum data storage – IBM 650
1960	PDP-1 for research use
1970	ARPANET
1974/75	First consumer computers
1981	Microsoft MS-Dos computer operating system
1981	Internet email protocol
1981	IBM PC
1984	Apple MacIntosh
1985	Microsoft Windows
1994	Netscape Mosaic web browser (beta version)
1994	Yahoo! web portal launched (incorporated 1995)
1996	Widespread use of teleconferencing. First web-based video-conferencing
1999	Hot-desking used as an option
2000	WAP mobile phones with Internet access, RIM Wireless email devices
2001	Post Sept 11 terrorist attack restrictions on travel in favour of teleconferencing

Source: *www.computerhistory.org and http://inventors.about.com/library/blcoindex.htm?PM=ss12_inventors*

new items. This epitomizes the virtualization of workplaces: it is not that certain parts of an organization have been outsourced to an online service company, but this new situation of virtually enhanced labour is often dealt with only from the point of view of the virtual and the virtual technologies which enable computer-mediated information systems. These databases have become comprehensive enough to warrant the label 'virtual environments' as they numerically simulate the current status of clients and of the business – their identities, preferences and purchasing patterns in tabulated form.

In Japan, Seven-Eleven convenience stores cater to lunch and overtime workers with not only refrigerated drinks but counters of freeze-dried soups and noodles which can be bought and microwaved on the spot. Using a point-of-sale system that monitors customer purchases, Seven-Eleven collects sales information for all its stores three times a day and processes this information to reorder stocks and give insights into changing consumer patterns. It is not just about how much is sold but when, and in what daily, weekly and annual rhythms and cycles. Satellite communications proved cheaper to use and less vulnerable to earthquakes than ground telephone lines, so every one of the over 6000 stores now has a satellite dish. Rather than promoting entirely new products, this type of system turns every store into a consumer sampling and surveying site, and allows changing demands to be catered to immediately. Quality control and pricing allow the stores to develop their own brands which will compete successfully in terms of price and style and which are more profitable. It can rapidly discern which goods or packaging appeal to customers.

> Seven-Eleven's merchandising and product-development capabilities are formidable. Its ability to sense new trends and churn out high-quality items is superior to other operators.
>
> (Michael Jacobs, Dresdner Kleinwort Wasserstein, cited in *The Economist*, 26 May 2001: 77–78)

The chain predicts daily trends: 'as customers become more fickle product cycles are shortening. Fashions in boxed lunches, riceballs and sandwiches, which make up almost half of a convenience store's daily sales', are especially short term in Japan, lasting about seven weeks. Fillings and sauces are changed to maximize sales. But each day, as temperatures rise and fall, hundreds of private local weather stations at stores anticipate whether demand will be for hot food or cooling drinks. Orders from stores are 'paperless'; electronically processed in less than seven minutes, they are delivered the same day from a network of distribution centres. Truck drivers carry identity cards which are scanned in when they arrive with deliveries – every trip and route is timed and performance reviewed: 'Seven-Eleven folk boast that their trucks run even more punctually than Japan's on-time buses' (*Economist*, 26 May 2001: 78).

Firms such as Wal-Mart use the Internet for product procurement on a global basis (*Economist*, 26 May 2001: 78; and see Chapter 5). Such e-commerce systems can be linked with other multimedia software to drill store attendants in company practices. There is a continual turnover of these workers who are part-time and who may never see each other, but rely on the consistent practice of co-workers to ensure that the store operates smoothly and continuously. In a country where people are still wary of using credit cards over the Internet, Seven-Eleven has increased traffic by turning shops into pick-up points where shoppers can pay in cash for purchases at online stores such as 'e-Shopping!Books' or Seven-Eleven's own site, *7dream.com*, and its 'branchless' bank IY Bank which allows customers to pay their utility bills. But analysts point out that 'only by using technology to expand markets or to reduce staff can a company earn back the cost of such an investment' (*Economist*, 26 May 2001: 78). One question is whether the Japanese focus on retaining staff and enhancing service will continue in the long Japanese recession since the late 1990s and early years of the twenty-first century.

Fewer writers consider the role workers play in this process of virtualization. Technological innovation, the daily performance of labour and the reorganization of the workplace have to be considered simultaneously in order to get a comprehensive view of what the process entails for work and workers. Work skills and the very nature of work and the workplace change with the computerization of organizations and their workplaces.

> Moving information back and forth between such realms – from the virtual to the physical, the verbal to the written, the personal to the mechanical – requires more than just technological protocols, especially when technological capabilities, message types and even the very norms of communication are themselves constantly changing. Vast numbers of individual workers, situated within simultaneously competing and cooperating institutions, are crucial to the daily maintenance and the gradual evolution of technological internetworks.
>
> (Downey, 2001: 220)

Yet in most studies of the Internet, virtual reality and expert systems make little mention of the labour involved. The virtual spaces of databases and online interactions are grounded in the social world of institutions, laws, culture and organizations. Those concerned with what will be labelled 'virtualized work' include those who *produce* the virtual as well as those who work through it:

- a wide range of skilled workers whose work has become more and more virtual in the process of it being computerized; and
- users and consumers of virtual environments including retail consumers, service professionals, and industrial and service firms (after Downey, 2001);
- technical and support workers who maintain and reproduce the technical and organizational networks which are the material supports of the virtual; and
- digital software 'agents'.

Managers of state and corporate institutions making regulatory policy and investment decisions will be dealt with in the following chapters along with firms proposing innovative services or business models which materialize the virtual.

Computer-mediated communications and the convergence of pre-existing networks allow new types of services and jobs, such as call-centre operators, to come into existence. Not only are these new positions, but they fit within the hierarchies and perceptions of the nature of the work, who should do it and what status it will have. This raises the question of labour markets and the availability of certain kinds of people, in certain places, to work.

> One hundred years ago, telephone operators were constructed as young women. . . . Today remote data-entry 'telework' is pitched to stay-at-home mothers while downtown data-design programming remains demographically male by a wide margin. Distinctions of gender, age, and mobility demand the inclusion of labour markets in the analysis – labour markets that are also grounded in space. . . . To the degree that people are the ones creating and populating the virtual spaces of information networks, the physical location of those networks depends in turn on where those people live and how easily they can get to their network-maintaining jobs. Thus where a network is grounded – whether suburban office part, an inner-city 'smart-building', or a free trade zone 'teleport' overseas – may have everything to do with who is there on the ground to operate it.
>
> (Downey, 2001: 233)

Call centres may be located far away from customers and from where products are manufactured. They depend on a host of qualities that the desired workers will have – punctuality, patience and willingness to work – but also on qualities that are place-based or regional, such as pleasant accents, telecommunications networks, and low wages and taxes on the type of business.

The frustration of computer crashes and losing one's work marks the moment at which computers betray the trust that must be placed in them. One must trust that the millivolt charges of memory banks, processors and magnetic storage media such as diskettes will hold and not degenerate into a meaningless jumble of zeros and ones, charged and uncharged coordinates in the gridworks of etched silicon. The fallibility of computers makes one question why it is that people are willing to invest so much in them. The sense of not being able to reconstruct creative work (an image drawn using a computer sketching program such as Corel Draw, or text written with a word processor) reflects the extent to which people exteriorize their psyche and realize that they have given over some vague part of themselves, of their sense of self, to the machine. Thus we can argue that workers themselves become somewhat virtualized, and that they experience this emotionally in their feeling of loss and frustration at computer crashes.

At the same time, even skilled trades and technicians often work entirely with computers. These form another class of virtual labourers: 'Unlike the popular image of a knowledge worker whose work is entirely symbolic, technicians also remain intimately involved with the material world. Technicians work routinely with machines, human bodies, and a host of other physical systems' (Whalley, 1997). Virtualized work is defined by the contradictions between its virtual, abstract and material aspects. Unlike the professions, *virtual work* is often not concerned with ideas but with the functioning of organizations, the operation of computer networks, the production of products and services. Compared to material work, work that may be primarily manual or involve the immediate performance of a service (e.g. chambermaids' work), in virtual work the relationship to material activity is truncated. Machines displace humans or accomplish tasks at a distance. Materials are no longer manipulated directly through the application of the force of one person or a team via hand tools. The activities performed may be under the guidance of other machines (computers) or other human agents (such as programmers or even a computer running an expert system which assesses alternatives based on a digitized archive of expert responses to a situation).

VIRTUALIZING SKILL: THE TRADES AND PROFESSIONS

Consider what is called a 'cybermation machine' in North America. This label describes a range of digital lathes, milling machines and metal stamping and cutting machines. Operated by software, the equipment interprets shop drawings of, for example, duct work. It can cut out with a cutting torch, stamp and even fold sheet metal with consistent precision. This replaces the traditional shop activities of sheet-metalworkers, which involved cutting out shapes using templates and assembling them. The collection of metal-cutting templates represented a solution to a problem which could be re-used, amounted the accumulated experience of the shop, above and beyond individual workers. Now, templates are software programs purchased from programming specialists. Other machines grind and mill precision parts according to parameters entered in their software. Workers who were once skilled operators of tools that performed only part of an overall task and which required continuous maintenance (oiling, sharpening, resetting and so on) have become loaders and unloaders of the machine and also computer operators (who deal more with computer crashes than with resetting, presses or lathes). The worker's relationship with cybermation machines (which become ever-larger and more powerful) shifts back and forth between the contradictory positions of 'hands-on' actual servant and 'hands-on-the-keyboard' virtual master.[1] One has to become comfortable with the virtual to work with a computer.

In virtualized work, manipulation of actual materials is shared with remote-controlled actions. Robots and software-driven machines and tools perform operations and are controlled via keyboards and joysticks from sound-proofed control posts.

Digital interfaces and typed commands replace analogue controls. Ironically, many companies have found that the best-selling digital devices are those that adopt the mimicry of analogue dials, the visible relationship one might see on a piece of equipment when turning a hand crank which screws in or out, up or down, a part of the machine. Skilled operators were those who understood the tolerances and had acquired an intuitive knowledge of the relationship between cranks, dials and levers, and the shifting centre of balance and centre of effort as force was applied to raw material.

The virtualization of work is not limited to machine shops. In principle, any skilled manipulation of materials from the simple activity of cutting a duplicate key to the complexity and teamwork of surgery may be accomplished via the mediation of robotic or remote-controlled machines. In some ways, being adept at video-gaming is becoming a good background skill for virtual work. For example, new methods such as keyhole surgery are done increasingly using remote-controlled robots which are able to work through smaller incisions which heal faster, allowing patients to recover in half to two-thirds of the time, and require fewer or no blood transfusions. Teaching surgeons may be heard to make such comments as:

> 'Surgery is becoming one big video game and [youth] they've got the skills. . . . All those parents who tell their kids to get away from the Nintendo may want to think twice,' comment American instructors such as Dr. Andrew Feldman, a Manhattan orthopaedic surgeon.
>
> (Thrush, 2001)

Newspapers report that cardiac surgeons who have taught over a hundred doctors how to use surgical robots argue that video-game skills are needed.[2] These skills include hand–eye coordination, fast reflexes and an ability to solve problems in a virtual reality environment (quoted in Thrush, 2001). Surgical robots consist of a set of operating 'arms' mounted on a tower or gantry over the patient. Interchangeable tools such as tweezer-like pincers allow the machine to mimic complex motions such as sewing sutures. They are controlled by a surgeon's hands at a console about the same size as an arcade video-game. Descriptions report a surgeon operating the device: '[He] slipped off his clogs to work a series of foot pedals, slid his hands into a pair of controllers and leaned forward to peer through the three dimensional video viewfinder', which displays images of the patient's internal anatomy picked up by video cameras whose miniature lenses are at the end of flexible fibre optic cables, allowing them to be inserted inside the incision itself. Coronary bypass and heart valve surgery is carried out using existing surgical robots.[3] However, the technology redefines not only skills but the role which surgeons play: 'The funny thing about this technology is that a lot of smart people can't do it and a lot of not-so-smart people are great at it', comments Robert Howe, a Harvard robotics engineering professor.

By contrast, 12- and 13-year-olds given a chance on the equipment can master complex tasks while older surgeons struggle to adjust to tying virtual sutures while a robot follows the motions physically. Thrush quotes one instructor: 'It's no big deal to the kids. . . . They've basically been doing this kind of thing all their lives on game and computers' (Nifong, cited in Thrush, 2001).

But it is naive to compare children and adults on the basis of skill without also comparing their differences in awareness of the implications of mistakes. Children do not have to worry about lawsuits arising from video-games or simulations. Surgeons may well be more nervous, and more cautious with robotic equipment because they are aware of the implications of mishaps and the control given over to software which assumes and is calibrated for standardized, 'normal' situations and bodies. Such comments, then, reveal an insidious romanticism of technology and an attitude among designers, trainers and boosters of such technologies that the ideal user is like a child. Users should abandon caution and surrender the cares and concerns of adult citizens. Furthermore, they should engage fully in the challenge of the technological puzzle or skill-set established by the machine without being distracted by context, implications or second-guessing the quality of a machine's output (see Users, below).

Indeed, technology does allow activities to be performed differently. Not only are specific tasks done from a distance; surgical teams, the administration of operations and follow-up care are reorganized. The blend of human and robotic surgeon raises questions of liability: are the limitations of the device a potential hazard to be assumed by the patient? The relationship between the patient and surgeon becomes a virtual one, because it is possible that they will never meet face to face. Instead, 'patient care consultants' blending subsets of nursing and psychology knowledge meet with the patients, ensure they are organized for the operation and follow up on their recovery.

One could imagine a gradual separation of the expertise of planning and coordinating complex surgical teamwork from the manipulation of surgical instruments. The 'hands-off' surgeon would play a role similar to the director of a movie, rather than also being a sort of camera-person as they are now. Such robotic systems also routinely archive the sequence of motions. While the machines do not experience muscle fatigue, such records may be used to guide interning surgeons on procedure. They will also allow the individual surgical procedures to be reviewed. They may provide evidence of malpractice, or even allow patients to take home souvenir diskettes of their operations!

ALIENATION

Archived in all its transistorized glory at the Computer Museum History Center, Mountain View, CA, sits SAGE, IBM's computer for the continental air defence system of the 1960s and 1970s. Scribbled in pencil on its aluminium controls is graffiti which gives a sense of the quality of work on the machine and the boredom of the technicians and military personnel who manned its screens: '"Don't you feel useless," and "I can't stand it," hint at the frustration of its operators' (Shadid, 2001b).

Work with the virtual imposes a distance from the actual. The effect of distancing people from their tasks is often accompanied by a shift in their skills from the manipulation of material objects to the weightless manipulation of digital representations. In this process, a 'virtualization' of skill-sets takes place. Where a metalworker was traditionally required to come to work with his own set of tools, the toolkit of the virtual metalworker is now as different as the skills required. For established workers, this shift is experienced as a loss of skill and pride in a proficiency obtained only after hours and years of practice, and as a new incompetence in front of tools and operating procedures which must be learned from manuals and in manufacturers' seminars rather than through the experiential approach of apprenticeships in which one 'sees and then does'.

Such a distancing can easily become an estrangement and alienation. The nineteenth-century critique of modernity and industrialization centred on two aspects of Victorian life:

1 The extraction of surplus value from labour organized in factories, which made an elite of capitalists fabulously wealthy on the backs of the Dickensian poor who lived in poverty in the working-class ghettos of the cities.
2 Ever-increasing rationalization through the imposition of timetables, school and workplace discipline, bureaucratization and economic rationality on to everyday life of all classes.

The expropriation of the value-added of labourers' work continues to take place when employers cream off the profit from the enterprises they own even though ownership structures have become more complex and diffuse. Karl Marx is best known for the diagnosis of the inequalities of capitalism and his political programme, *The Communist Manifesto*. This was based on a less well-known critique of alienation. He sought to reveal the actual losses by workers through an economic analysis. This was the price paid by entire populations for modernization. More insidious, his sociological analysis suggested that workers, in exchanging the possession of their work for a daily wage, gave up something of the pride in their work; their sense of the fruits of their own labour as a realization of themselves. For Marx, labour was a process of giving material form to human ideas and spirit – the defining mark of Marx's humanistic vision of *homo faber*. In the categories of our analysis, this is a process of actualizing the virtual and the abstract.

What are the implications of virtual work? This labour operates between the abstract, the virtual and the actual, but privileges the symbolic: manipulating signs, mailing text, programming code, data entry in spreadsheets or even maintaining computer equipment for the overriding purpose of this work with signs and images (work on hardware is very much 'material' work, but very few technicians work solely with hardware except for specific groups, namely the lowest level of assembly-line workers, packers, installers and equipment salespeople).

THE QUALITY OF WORK: DATA OVERLOAD AND INSTRUCTION RITUALS

Virtual technologies generate an explosion of data because they make not only material but virtual forms of data and information accessible. I understand these as disordered and ordered forms of representation of coding, respectively. Some may encourage workers to look around at who they are working with, to talk to others to create a workplace community, and to participate in any training programmes offered (Eyerman, 2001; Kirk, 2001). But the result is more often anxiety and impatience which permeates not only online but telephone and face-to-face conversations.

'The great Information Age is really an explosion of non-information; it's an explosion of data. To deal with the increasing onslaught of data it is imperative to distinguish between the two; information is that which leads to understanding' (Schachter, 2001). This overwhelming amount of disorganized and often contradictory data which people experience partly resembles the experience of listening to a national weather report going region by region, where one is distracted by a phone call by the time the announcer reaches one's own region and has to make do for the rest of the day on the basis of what one surmises the weather should be, based on how it is on the far side of the country. One often has to generate knowledge, or understanding, by marrying incomplete or even inappropriate information with other data which puts everything in context that one has to pick up on one's own.

Under the stress of the increasing need to synthesize new knowledge and understanding, and to organize data into informative patterns and relationships (i.e. into information), people are less patient and forgiving of how others convey information to them. Not only do they want to get to the point, but they dispense with the small talk of the past which often carried a level of redundancy but which allowed misunderstandings to be corrected before anyone started work. It was chit-chat that communicated the *why* of requests, allowing people to understand what they were being asked to deliver, what its operating environment would be and what uses it would be put to. Thus tolerances and operating parameters were understood and suggestions made in conversation.

One of the great impacts of the expansion of the field of information by digital virtuality is this dehumanization of social relations, especially in the sphere of work. This is manifest in the shifting discursive environment of workplaces. 'Get to the point' and 'Get on with it' cut short not only discussion but opportunities to get to know and to update our understanding of each other's concerns and engagement with the challenges and opportunities of everyday life. We commiserate and support each other less, and forgo the micro-opportunities to celebrate each other's achievements. For everyone at work, the time spent discussing the meaning of information is key to disseminating knowledge and reaching a consensual vision of goals which are in turn the foundation of successful organizations. Social networks supply information on more than just work, allowing workers to adapt to changes in communities and to changes in organizations and employment patterns.

Without chat, everything in an exchange must be thought through ahead of time and made explicit – rather like leaving a phone message. Some argue that every conversation should be treated as an instruction. Warman, for example, compares specific task- and goal-based instruction. In the latter, the receiver will need an overall goal stated up front, as they will immediately begin to formulate a plan during the course of the instruction. Six elements are important for achieving understanding when giving instructions:

- the reason for the instruction;
- the objective;
- what procedures are expected;
- the time allowed;
- what may be expected to happen during the task or mission;
- how errors can be recognized so that the instruction-taker will know when to regroup or return for more instructions.

(Warman, cited in Schachter, 2001)

The problem with this approach is that it assumes that all knowledge or understanding is held by the instruction-giver, and that all information is also centralized by this key figure. In reality, managers depend on having *knowledgeable* employees who possess their own capacity to synthesize information given in instructions with their own experience and networks of information. Furthermore, successful organizations are those that nurture the two-way flow of information, which is to say that instruction-givers look for feedback from instruction-takers. Manager and subordinates engage in carefully structured rituals at 'work' in which they trade positions back and forth. These *instruction rituals* take the form of 'meetings' in which who speaks, in what order and what gets discussed is carefully crafted. These are the social media of instruction and of information exchange between people at work. These rituals, who may engage in them and what information and instruction may be exchanged are usually obscured by the core content of an organizations 'culture'.

USERS

> 'Back in the early days of computing, machines were very large, they were in glass rooms. In order to use them, people delivered decks of cards, punched cards, with their programs and data, and somebody else took those cards and ran the program through the machine and gave you the results. You never touched the computer. And Digital basically went into the business of providing machines where the user actually sat at the keyboard, typed his program, you know, did the work.' (Ed Kramer, past Digital senior VP in an interview conducted by the Computer Museum History Center, Mountain View, CA).
>
> (quoted in Shadid, 2001a)

Virtual workspaces depend heavily on digital technology and computer-mediated communications. Every participant in a shared collaborative workspace requires their own access device, whether it is a handheld wireless device, a desktop computer or a room set up for video-conferencing. Before the dot.com bubble of the 1990s, Digital Equipment set the path followed by all dot.coms as they rose and fell. Digital took the first step towards the PC with its idea that every researcher should have their own terminal. Their first technological breakthrough was PDP-1 in 1960, an alternative to the IBM mainframe that was the first to come close to being interactive.[4] The 'programmable data processor' (PDP) was the first of a line from the commercial revolution of the 1965 PDP-8, the first mini-computer (smaller than a room) to their line of VAX computers which powered engineering, census, accounting and medical computing even into the 1980s. These terminals were the first technological changes by which everyday work began to become more and more closely interconnected with the virtual. This first took the form of accounting databases and quickly expanded with the creation of text messaging and email which allowed forms of textual communication. 'It was the first step toward personal computing . . . but it never foresaw a person in every home having a computer' comments Kramer.

By contrast to the North American or European business use of computers, in western China (the poorer areas) in 2001, PCs and their keyboards sport elaborate velvet covers. Most schools could never dream of providing such equipment for students, but the poshest, experimental 'key schools' offer labs in which one removes one's shoes and sometimes must even wear plastic slippers or booties to use IBM-type personal computers. The reverence for the machines, the special labs and coverings separate them from the 'hands-on' world of everyday work and study – not to mention play.

Most of these computers appear to be used to teach typing skills, English and word-processing. Some places use them for teacher training and for distance education. Adult learners can go to a local lab to study. Some schools have access to the Internet, but it is heavily censored. A similar aura of inaccessibility surrounded the IBM mainframe computers of the 1960s and 1970s. Programmers were like a priesthood. The shift from specialist operators to opening the computer up to generalist *users* marked the first step in making digital virtuality accessible to a wide population. The Digital approach, which provided a terminal and keyboard for users to do their own data entry, was quickly copied by IBM, Honeywell and other manufacturers. In Euro-American societies, a revolution has occurred in attitudes towards computers, which have been integrated into the everyday equipment of workplaces, schools and, increasingly, into the home.

> 'We were building kind of the first personal computers, the first computer that researchers, engineers, scientists, people in the biological sciences could use,' comments Grant Saviers, a former Digital VP, 'This really changed the world'.
>
> (quoted in Shadid, 2001a)

However, the PDP required mastery of proprietary software. Digital understood the hardware side of the PC revolution, but misunderstood the basic qualities of the computer as what even Turing called a 'universal machine'. Software independent of specific machines (such as UNIX or DOS) was dismissed, eventually leaving Digital far behind as open-source software became standard and as the computing power of PCs allowed programs written for Microsoft's near-universal MS-DOS to compete with refrigerator-sized computers. In fact, '"Digital had a dim view of non-technical people using computers," says Kramer, "It built sophisticated tools for sophisticated users"' (quoted in Shadid, 2001a).

Digital's approach required that scientists using research programs, and administrators using accounting software, learn sufficient computer skills to make a transition from merely interpreting tables, graphs and accounts to opening files of the digital data, entering and selecting information and mastering the acronyms and codes which prompted the computer to make a specific calculation. This was the first step towards changing the nature of their work and the architecture of workspaces.

COMPUTERIZATION OF THE WORKPLACE

The computerization of workplaces is reflected not only in the incorporation of digital virtualities such as online documents and computer-mediated telecommunications, but in the forms and postures of physical work. The history of the desk and filing systems follows the movement into the virtual, from the nineteenth-century roll-top desk with its pigeon-holes for filing, to the desk and drawer designed for industrially produced standard-sized papers to the desktop displayed on computer screens and shared online workspaces and files such as is offered by companies such as Yahoogroups or Groove.com.

Beginning in research labs, computerization of the workplace meant that people had to work in front of, first, Digital's PDP terminals (above) and later smarter personal computers which functioned as terminals for Windows-based programs and online databases. Beginning with research workers and lab technicians, work moved on-screen rather than on a table top or lab bench. In the office, this was fundamentally to change clerical work.

It would be a mistake to assume that the form, placement and relationship to other business equipment – and to the bodies of users – was simply determined by the requirements of technology. In hindsight, early desktop computers seem bulky, as if designed for the convenience of technicians or assemblers rather than with regard to desktop 'real estate'.

> Computer design and work place are far from immutable. In fact, they are utterly negotiable. Both currently reflect the prerogatives of generating capital and controlling the networks and hierarchies of work.
>
> (Hayes, 1995: 178)

Bulky personal computers took over workers' desks and 'Windows' graphical interface or 'the desktop' took over from the blotting-paper. Office furniture manufacturers recognized these changes, developing workstations organized around the computer, screen and keyboard and their thirst for power outlets and tangles of cabling. Eventually a place was even found for the computer mouse. The long hours 'at the keyboard' and in front of the screen entailed height-adjustable tables with multiple levels. The 'keyboard tray' replaced the middle drawer of the Western office desk, while the computer screen blocked the view from desk to desk in offices. The typewriter table all but vanished with its namesake.

But the computerization of the workplace is more than simply the virtualization of tasks and business interactions (for example, rather than telephoning or meeting face to face, one might email a person or even use a web-based video-conferencing connection). Others have argued that computerization has implications for the types of analytical thought, the logical and linguistic skills required of workers. Rather than critical, bottom-line analysis, an awareness of the political economy or the ethics of decisions and actions,

> The idea that the capitalist system wants a good many critical thinkers is simply absurd – it can only spell trouble. . . . Thus the point is to produce the human as puzzle-solver, not really as critical thinker. . . . The mind is thus habituated to thinking only in limited, even if complex, ways.
>
> The beauty of the computer is not simply the speed with which it computes, nor even all the troublesome work-resistant workers it can replace, but that it can simultaneously, powerfully shape the mind and the personality. Thus, if *successful*, computerization will enable the production of the human as computer.
>
> (Neill, 1995: 192)

In effect, a virtual reality of sub-political, technical concerns replaces the 'messiness' and broadband complexity of the world of human interactions. Rather than having to translate orders into terms which may be understood by workers, or to accommodate their physical frailties, command and control at a distance allows decision-makers to dissociate themselves more fully from the concrete implications of their decisions and the organizational frameworks they set in place. Is this like the F-16 pilot, for whom dropping a laser-guided bomb is a bloodless but exciting video-game-like exercise of responding to a digitalized, abstract '*threat*' on a cockpit screen at a distance of 100 kilometres rather than a directly encountered, concrete *danger* (see Chapter 6)?

THE 'WORKSTATION'

Home and office computing required easy-to-use computers geared towards non-engineers who were not interested in writing their own programs or being trained to use one single type of software which would only ever run on one manufacturer's computers. The Internet ushered in an era in which data was more and more mobile. Rather than being entered by a single researcher to solve a single set of equations, databases held in memory banks had to be capable of being accessed by distant machines made by other manufacturers, and of being moved from computer to computer.

Two periods may be distinguished in this shift. From the introduction of the terminal and desktop personal computer in the mid-1970s until the end of the 1990s, computers were fixed with the workspace reorganized around them. Occupying large amounts of desk space the screen and computer at first sits in the centre of the desk with all other materials such as pen and paper relocated to one side or the other. Where personal interaction is important or reading paper materials continues to dominate, the computer sits to one side, its display unit angled at 45 degrees to the user and swivelled from time to time so that clients can see its screen. Space allocated to the computer moves to the corner of an L-shaped or curved desktop. Sitting in the corner of an office cubicle or room, the view and spaces behind the desk are more accessible, but in office cubicles one works with one's back towards visitors in a fundamentally insecure position; workers use the reflections on the screen to spot arrivals. Holes in the top of the desk may organize computer cables and power cords, but limit the ability of the user to relocate the screen, computer 'box' or CPU or the keyboard. Often the cables are of limited length and screwed into the 'ports' or jacks of each component – again, not designed to allow easy disconnection, movement of the computer and reconnection.

On the scale of the office, work is more fixed to locations than ever, defined by each worker's computer. The cubicles form a fixed architecture of work expressed in terms of space allocations rather than the area necessary for changing tasks during the course of the work day or project cycle. Despite the 'open concept' and corporate office environment with its token plants and office dividers, the logic of the workplace is dictated by the increasing computerization of work. Against the isolation of workers in front of their individual computers, online systems develop to allow computer-mediated communications such as real-time interactive chat. If email was thought of as leading to a 'paperless office', so 'shared virtual workspaces' and 'shared files' extend the logic of an office of monadic individuals whose communication is narrowed to the necessary and work-focused channels.

THE FLEXIBLE OFFICE

The fixed position of work and the lack of adjustability of early arrangements of computers and terminals, the repetitive nature of typing and the long hours of fixed focusing on a video display screen required that the human body adjust to the technology. Enticed by narratives of progress, employees were induced to submit to the discipline of immobility. The result was a rising number of claims of repetitive strain injury (see below), panics over radiation from the screens and worries about the long-term effects of being surrounded by fields of electric current.

By the turn of the twenty-first century, a new attitude is evident which acknowledges the possibility of flexible placement and changing location of the computer. Desktop space is freed up by allowing the CPU to be placed underneath the desk, and slightly longer cables are provided by, for example, screen manufacturers. Improving display technology and portable computers offer the potential to become 'desktop replacement' machines, which proliferate along with docking stations integrating them with the older pattern of fixed screen and keyboard locations.

The shift towards a flexible workplace responds to the proliferation of teamwork on projects and anticipates a gradual movement towards deskless computing on the basis of palm-sized computers or terminals. These, for example, permit wireless access to digital virtual environments at any time, allowing them to be used anywhere – much as one might read a book on a bus or in a café, not just in a library. An example is a combination of bar-code readers and wireless terminals which will also record signatures on their touch-sensitive screens such as those used by workers in delivery companies. These allow customers to sign for parcels and also report the exact time of delivery.

A second form in which computerized work becomes less fixed in place is the ubiquity of integrated computer control of devices from refrigerators to lathes. Wired to the Internet, designers and information architects dreamed of a time when their status could be accessed remotely via small communication devices. Mobile pagers and information devices by firms such as Research in Motion (RIM) exploited digital telephone technology to provide handheld access to email and to shared files formatted for the World Wide Web. Digital phones were also equipped to browse the Web.

The flexible workplace is typified by mobile office furniture and arrangements which are changeable to at least some degree based on workers' patterns of cooperation on assignments or their individual tasks. This requires a separation of furniture such as desks from walls or dividers. Wireless connections (infrared and high-frequency radio signals) may replace cables except in low fault-tolerant situations where large amounts of bandwidth are required (e.g. audio-video-processing). The very term 'flexible' responded to charges that the inflexibility of computer gear and seating caused strains and injuries. This ergonomic concern is a final feature of the flexible office. Ergonomics consultants – not the old science of human ergonomics and biometrics – emerged as a new group of support workers created by clerical and administrative computing arrangements.

In short, the changing nature of work tasks and the role which workplace design played in enabling those tasks and solving work problems may be seen in the changing arrangements of offices. These include:

- a shift from offices laid out around fixed individual 'work-stations' and cubicles towards space allocations of more flexible design;
- changes in the design of office furniture and equipment;
- the emergence of virtual environments of shared digital files and databases;
- increasing ergonomic concerns over strains and injuries.

UBIQUITOUS COMPUTING

The vision of wireless data transmission systems and multiple and modest but interconnected computers in all appliances and artefacts offers the possibility of an all-embracing virtual environment of data and communication layered on to the material world and spaces of everyday life. Chips could be used not only in one's car but in every device from a 'smart lightbulb' which would adjust its luminosity in response to sunlight to chips implanted in bodies which would communicate those bodies' preferences (sending signals to turn on specific lights at a certain level when a specific – and authorized – user enters a room). The goal would be to put this in the background of attention. Devices would maintain an ambient environment of information and computer-mediated functions which would change as users moved from place to place. Rather than a disembedded virtual reality, the focus is on the interrelationship of abstract data, virtual telecommunications spaces and material or embodied interaction.

> [Current] information technology is more often the enemy of calm. Pagers, cell phones, news-services, the World-Wide-Web, email, TV, and radio bombard us frenetically. . . . But some technology does lead to true calm and comfort. There is no less technology involved in a comfortable pair of shoes, in a fine

> writing pen, or in delivering the New York Times on a Sunday morning, than in a home PC. Why is one often enraging, the others frequently encalming [*sic*]? We believe the difference is in how they engage our attention. Calm technology engages both the *centre* and the *periphery* of our attention, and in fact moves back and forth between the two.
>
> (Weiser and Seely Brown, 1996: 4–5, cited in Galloway, 2002: n.p.)

The discourse of ubiquitous computing offers the illusion that it is possible to 'domesticate' the computer without users being 'domesticated' (Galloway, 2002). However, it is likely that by including a user's context and perhaps direction and speed in the automatic calculation of what information is relevant and the adjustment of some environments to the user's personal preferences (choice of background music, ambient temperature, etc.), it is more likely that contextual advertisements, advice on routes and opportunities 'coming up' on one's path will precisely attempt to domesticate the user, to lead people as much as possible to commodified exchanges and experiences.

If enticements to entertainment can be embedded in everyday life, so can work. Via its wireless networks, ubiquitous computing implies further erosion of the spatial and temporal separation of private and public life, home and work. For example, one might always know where one's co-workers are in case they need to be contacted. These divisions would need to be reinforced in other ways such as explicit cultural norms on 'work time' or the working day and family time or time off. The question of the workplace is raised again. However, these will likely be seen as valuable contexts for collective projects. Workplaces may be thought of as merely a useful tool for the social aspects of work. However, their importance for the coordination of material goods, and more abstract knowledge and routines, cannot be underestimated.

CLERICAL WORKERS

The computerization of the workplace and virtualization of work described above had perhaps the greatest impact on clerical and secretarial workers, a job that was all but eliminated in the process. The spread of computers to almost every person in the workplace resulted in a democratization of data input and typing in many organizations. Although large organizations continued to maintain some back offices dedicated entirely to data input, the task of keying in information (such as responses to surveys or customer payments) has also been widely automated. In the past, secretaries were responsible for the filing, typing, proofreading and formatting of documents. Nowadays, typewriters are scarcely to be found. Secretaries undertook many of the coordination tasks (such as establishing meetings and message-taking) between analysts and between decision-makers. Currently, secretaries are highly unusual, apart from quasi-professional medical and legal secretaries and the 'executive assistant'. For a brief period, dedicated word-processing machines were operated by specially trained staff. However, the rise of the personal computer in front of every worker meant that each worker became responsible for typing in and printing out information and documents. The coordination of work was undertaken less through the telephone, in real time, and more and more through email. Typing, once a 'pink-collar' activity, became a required skill not so much taught as self-taught, as children learned to operate home computers to browse the Net at home.

The dissemination of keyboard activities such as typing one's own documents and the use of email is perhaps one of the most noticeable changes, not only in 'office' work. Keyboard-equipped computers may be observed from mechanics' shops to retail businesses and professional offices. This was not an optional change in work; it has been universal and took place over less than fifteen years from the introduction of the first desktop personal computers such as the IBM PC. It is now inconceivable that an employee would not know how to operate common programs on an IBM PC-type computer. Yet business computing has become more complex involving the retrieval, comparison and computer processing of information from different databases. Email systems have developed into personal information managers and agenda-setting programs which are capable of displaying one's availability for meetings to co-workers. Word-processing software has metamorphosized into desktop publishing systems which integrate graphics and the spreadsheets produced by software bundled as 'office suites'. The graphics require mastery of still further software to generate images which are entirely digital and virtual or scanners to get hard-copy materials into the digital–virtual realm of computer media. The documents may be printed out or published in various electronic formats for dissemination online.

Only certain areas, such as webpage publishing, have seen the return of new forms of clerical work, but in the guise of graphic design and public relations. The dream of simply being able to 'print' documents to the Web without serious formatting difficulties occurring has not happened for most business users. The complexity of business computing means that a worker must now be a jack-of-all-trades. Systems such as project databases or customized annual account reporting programs will be used infrequently, meaning that every time a worker works with the program, they approach it as neophytes, having forgotten the last session six months or a year previous. These inefficiencies and the growing demand on workers to learn and upgrade their abilities and competence as 'good users' highlight other inefficiencies of the digital virtualization of work.

By far the greatest demand placed on office workers was the reorganization of tasks around desktop computers. The effect was to combine tasks which were once separate – for example, filing, planning and analysis. The result was new concerns over the embodied nature of office and service-sector work.

> The relationship between repetitive strain industries and IT is a special one. The primacy of computer technology in the

> workplace has been a great leveler, affecting managers and managed, factory and office workers. By combining jobs, clustering work tasks, and monitoring performance, business firms have used information technology to radically revise the way nearly every employee works. The problem is that computers, more than any single previous technology have funneled work tasks into a very narrow range of physical motion.
>
> (Hayes, 1995: 177–8)

BODIES AT WORK

Every workplace element makes physiological demands on the user's body. The fixed angles and design of components such as keyboards were associated with a surge of complaints alleging repetitive strain injuries (RSIs). In the early 1990s there was much discussion of Carpal Tunnel Syndrome, affecting the wrists and hands. For example, RSI accounted for 56 per cent of 331,600 graduate-onset work-related illnesses tracked by the US Department of Labor's Occupational Safety and Health Administration (OSHA) in 1992. 'Damage to wrists and hands is now one of the fastest growing and most widespread occupational hazards.' RSI in the form of related workers' compensation and absenteeism now costs corporate America $20 billion a year. Given that more than 50 million people labour at office computers in the USA, these claims were projected to grow (Barge, 1994; Sambyal, 2000: 71). By 2000, Sambyal reported, many manufacturers faced lawsuits. One against Kodak charged that the keyboards on its computer system caused injuries because of its height which required users – writers and editors – to hold their hands at an awkward 45-degree angle to type. 'Among the generally accepted culprits are poorly designed workstations with keyboards placed too high, ill-fitting chairs, stressful conditions, and extended hours of typing' (Sambyal, 2000: 71). Pinsky cites the US National Institute of Safety and Health to argue:

> Automation has been successful in shifting the locus of work from the level of the trunk to the upper extremities [arms]. The workloads are now lighter, but the work pace has been increased. As a result, the associated work forces are concentrated on smaller parts of the anatomy, i.e., the ligaments, tendons, muscles and nerves that control the hands, wrists and arms of a worker.
>
> (Vern Putz-Anderson/NIOSH, 1988, cited in Pinsky, 1993: 12–13)

However, establishing a cause-and-effect relationship between keyboard use or other input devices and Carpal Tunnel Syndrome or other RSIs has been difficult. On the one hand, an industry arose of office ergonomics consultants and manufacturers of ergonomically designed office furniture whose curves were styled to visually emphasize its supportive qualities. On the other hand, computer users' complaints first increased and then tended to level off by the end of the decade. In 2001, the Mayo Clinic reported surprise in their study of computer use and Carpal Tunnel Syndrome: heavy computer use is not correlated with risk of developing Carpal Tunnel Syndrome. One of their researchers, J. Clark Stevens, commented:

> We had expected to find a much higher incidence of carpal tunnel syndrome in the heavy computer users in our study because it is a commonly held belief that computer use causes carpal tunnel syndrome. . . . The findings are contradictory to popular belief but nobody has studied the problem carefully.
>
> (Stevens *et al.*, 2001)

Other researchers and commentators such as Edward Shorter of the History of Medicine programme at the University of Toronto dismissed Carpal Tunnel Syndrome by comparing it with hysteria. Earlier crippling illnesses such as chronic fatigue were believed to be caused by one thing or another, but no link could be found. These ailments are real but ideal, *virtual illnesses*. These outbreaks are often dubbed 'public hysteria' by the medical community. They generate actual symptoms but no material cause or consistent physiological deficiency can be found. Virtual illnesses, however, are different from mass hypochondria in which people fear they may have an actual illness which has not only symptoms but causes and underlying physiological patterns such as lowered levels of blood sugar or enzymes. This confusion suggests that medical professionals would benefit from a better understanding of the distinction between the virtual and the material. Newspapers quoted Shorter as saying:

> 'Even before chronic fatigue there was concern about hypoglycemia in the 1950s, and chronic brucellosis in the 1930s. These are all media-spread illness attributions.'
>
> 'In the 1950s, there were people going around with packages of Hostess Twinkles in their briefcases so they could have a brief glucose top-up.' Mr. Shorter says even chronic fatigue is 'starting to go the way of the dodo bird'.
>
> (quoted in *Citizen*, 2001: A12)

Stress, for example, is very difficult to pinpoint materially. It comes from a variety of sources and different individuals react to it in different ways. Stress-related absences are an important factor in absenteeism, and in many places are an accepted type of sick leave. The immateriality and intangibility of stress make it a good example of virtual illness. A similar situation prevailed in the debate over Gulf War Syndrome. Was it a psychosomatic post-traumatic illness, or a condition materially caused by something such as exposure to shells hardened with depleted uranium or exposure to some sort of battlefield poison? Without a common manifestation and a causal relationship to a material contaminant, Gulf War Syndrome remains a virtual disease with veterans dying or dead of their material symptoms but with neither a coherent enough symptomology nor an underlying cause for governments and health practitioners to acknowledge the syndrome as an actual ailment. None the less, one empathizes with the general anxiety around computerization. Even if RSI is not correlated directly with computer use, the question merely shifts to a more ideal level of virtual insecurity and abstract threats (see Chapter 8). Hayes laments:

> how misinformed we are about information technology's real-world effects and . . . its threats to our personal health and economic well being . . . [raises] disturbing new questions about information technology. Our answers need not be constrained by predilections to turn back the clock or to make the best of what seems inevitable.
>
> (Hayes, 1989: 179)

TECHNICIANS AND SUPPORT WORKERS

Virtual work is often technical. A number of authors have pointed to the importance of technicians as mediators between parts of larger systems and between systems such as communication networks and organizations or even patrons. Examples might include telephone operators in the first half of the twentieth century or help-desk technicians in the latter half. 'With many of today's newest service jobs dependent on Internet work . . . new ethnographic studies of technical work have begun to emerge.' Barley and Orr argue that this kind of labour, defined as technical work in many studies and as 'boundary work' at the interface between interconnecting networks or 'internetworks' by Downey, has four main characteristics to which we will add a fifth:

1 complex technology is central;
2 contextual knowledge and skill are both necessary;
3 theoretical and abstract knowledge are also necessary;
4 a 'community of practice' exists, serving as a repository for all this knowledge and skill.

(Barley and Orr, 1997: 12, cited in Downey, 2001: 228)

5 translation back and forth between the virtual, abstract and material is necessary.

'Support people occupy the boundary between the known and unknown, between software that works and software that does not' (Pentland, 1997). As anyone with a computer at home or work knows, technicians 'link us to technologies that are nearly transparent when they work and troublesomely opaque when they do not' (Barley and Orr, 1997: 14).

> No matter what automated protocols are in place at any given moment, they will be imperfect and incomplete; disparate information networks can only work together through the efforts of specific workers who maintain the links, transform the content and police the boundaries between those [interlocking] networks . . . smoothing the transition from one network to another.
>
> (Downey, 2001: 225)

In effect, support workers such as technicians and those who answer troubleshooting hot lines negotiate the boundary between the virtual and the material world. Their labour comes to most people's attention only when it fails and the weightlessness of the Internet crashes to earth. One of the understudied aspects of such work is its self-effacing nature which contributes towards mystifying the dependence of digital virtuality on material technology, organizations and humans. The details and material conditions by which the virtual has been digitally brought into everyday life are concealed. New categories of technical worker have been created to take advantage of telecommunications and computer-mediated communications at the same time as assisting in their use. Help-desk and call-centre workers are new occupations which rely on virtual communication technologies and occupy a specific *virtual labour niche* closely related to technicians and service operators. There are estimated to be between 250,000 and 300,000 help-desk workers in the USA (Barker and Christensen, 1998; Leonhardt,

2000). These workers are low paid and often part-time, guided by expert systems which take them step by step through a diagnosis or troubleshooting process. As a result they are highly vulnerable to lay-offs and to poor working environments.

The new economy creates an 'hour glass labour market. . . . At the high end are the secure, well-paid jobs of professionals and high-tech positions, and at the low end are the poorly paid, vulnerable service jobs. The jobs in the middle have evaporated' (Ayerman, cited in Kirk, 2001: H7). Office work, particularly for women, has been degraded into dead-end positions and hourly wages have dropped. Clerical workers whose jobs were eliminated in the 1980s moved into customer services, which in turn has been transformed into competitive telephone sales and customer complaints positions over the 1990s. Finally, these jobs have been outsourced to telemarketing firms which serve several firms. These workers are

> simultaneously hidden and revealed, often physically located in a remote site – perhaps one with lower office rents, or closer to particular labour markets (even overseas) – yet virtually the first point of contact between company and consumer. Support workers are housed together in carefully crafted work environments, surrounded by sophisticated technology, and expected to work at a rapid pace to process calls, in an industrial factory-floor environment, yet they remain service workers. And the entire purpose of employing legions of support workers, immediately accessible to customers either by phone or by email, is *to use a sort of virtual transport to substitute for expensive on-site service calls.*
>
> (Downey, 2001: 231; italics added)

The virtuality of their labour, its quality of being a form of magic carpet or virtual transport, also allows the help desk and the organizational architecture of the call centre to be generalized to other occupations which involve onsite visits. Thus welfare agencies have attempted to virtualize some aspects of social work and public health work, even parole supervision in the form of call-centre work. The telephone substitutes for some visits (although it is generally impossible to eliminate all onsite instruction and follow-up). In other cases, new services such as 'hot lines' are created. An example is a service called 'Mother-Risk', a mother-and-child medical hot line provided by Ontario hospitals as a free service that can be consulted by pregnant women and parents of infants. On the one hand, the service provides expert answers to minor worries, effectively eliciting questions that would normally be asked of friends or family or older women who have experienced pregnancy, for example. However, the service uses the telephone as a consultation medium to help mothers decide whether they should visit clinics or hospital emergency services. 'Mother-Risk' aims to divert information requests and non-urgent cases to reduce the use of outpatient and emergency services.

Such services also provide webpages of answers to frequently asked questions (faqs) and invite queries by email, but these are often less satisfying and not reassuring, because the material bases on which a parent might make a judgement about trust in a medical practitioner are absent. For example, it is impossible to engage in the sort of responsive ballet of body positions and non-verbal cues by which empathy is achieved, and by which we assure ourselves that our concern and its basis in our values has been understood. Statistics on questions, answers and the user's responses to demographic queries allow a kind of surveillance function by which health administrators may chart changing concerns and trends in observed symptoms. It lays the basis for attempts to discipline parental anxiety, infant and foetal illness and risks. But the technicians' problem is always that satisfactory answers generally include more than merely technical information (about a technology, such as the steps in using an automatic bank machine) but also wider wisdom about how to avoid the problem in the future (how to tell if a machine is functioning, or how to ensure a magnetic access card is not demagnetized and so on). But to offer this answer requires that the support worker is supplied with information about contingent aspects of a problem and the context in which a problem occurred – sometimes embarrassing information ('you did *what* with the card?!') which often requires trust and a knowledge of the web of other activities in which use of a technology (or perhaps something more urgent, such as feeding a baby) is embedded.

It is expected that telework and the dispersal of work to a variety of lower waged areas will offer protection from potential attacks and disruptions in major cities, in the wake of the terrorist attacks of 2001. In Canada, for example, a country with a population of only about 30 million, Statistics Canada reports that there are almost a million teleworkers, up 40 per cent from six years ago, with over 50 per cent of workers interested in working from home to avoid typical sixty- to ninety-minute commutes. Despite the loneliness and challenge of keeping time for home life, telework is touted with improving the extent to which workers can manage their own time more productively. For example, according to one survey, American Express telecommuters handled 26 per cent more calls and produced 43 per cent more business than did their office-based counterparts. Compaq Computer found productivity increased from 15 to 45 per cent (Canadian Telework Association survey, cited in McNair, 2001). However, telework figures include work on the road (in the USA, 24.1 per cent) such as inspectors, salespeople and site supervisors as well as those who work from home (21.7 per cent) and those who work at telework centres (7.5 per cent) and from satellite offices (only 4.2 per cent).[5]

Finally, these figures vary in weight from country to country depending on using multilocational forms of telework, via telephone or wireless and regional satellite offices. Although figures have been assembled for 1999, the mixture of forms of telework suggests caution in comparing figures between countries and surveys, and hints at the likelihood that quite different forms of work and conditions are being conflated in the surveys. Even if treated as approximations, the wide differences in the numbers given by different sources suggest that the figures available from sources concerning telework and even concerning work over the Internet reflect their importance to advertising and promotional campaigns for information and communication companies and telecommuting boosters (Table 6.2).

Table 6.2 Teleworkers (from home and from remote locations) as approximate percentages. Sample figures for various workforces

Countries	*1999 aggregate**	*2000–2001 home****	*2000–2001 on the road****
Germany	n/a	2	6
UK	5.5	3	14
Canada	n/a	8**	n/a
USA	12.5**	21.7	24.1

Source: *ETD, 1999; **Statistics Canada, 2000; ***Emergence Employer surveys, 2001

Telework involves not only the virtualization of work but a stark shift in the time and space in which work is done. There has been little public debate about telework, and where there has been discussion there is no mention of the challenges that its virtual qualities add to both managing and resisting moves to press employees to increase their pace or expand their work. Menzies is quoted online:

> People enter into the discourse only in terms of that economy – as new skill-sets needed, or as redundancies to be adjusted into retraining or workfare programs on the margins. Furthermore, the discussion has focussed almost exclusively on the state as the agent to manage the social-adjustment aspects of the restructuring agenda and to mitigate any untoward effects. Government bureaucrats, in consultation with experts from business and labour, are qualified to script what's to be done, and to do it. The rest of us are bystanders.
>
> (Menzies, 1998)

Although there are some merits regarding the opportunity to manage their own time, widespread experience suggests that workers face a challenge in being compensated fairly for overhead costs and risk entering a job ghetto. Time spent working overflows into 'family time' and can become all-consuming. In effect, the home becomes host to a virtual workspace both online and in the physical presence of the home office or workspace. The blurring of work and home life introduces conflicts over time and over the computer-mediated presence of virtuality into the heart of family dynamics.

Most of the debate has focused on the experience of individuals, but the statistics indicate the massive numbers involved. In North America, for example, various forms of teleworking are rising overall at a rate of between 15 and 20 per cent annually. Because it is computer-mediated, the detail and productivity of teleworkers can be much more closely measured. The work content and patterns themselves are an outgrowth of the corporate restructuring of the 1980s and 1990s.

Optical fibre and high-speed national networks are the state's contribution to facilitating the process. For those who telework 'on the road' these systems allowed reports and information to be filed with head offices much more quickly. However, office-based intermediaries and other employees who were required to liaise with those on business travel were displaced. These concerns only summarize the challenges of coming to terms with the virtualization of work.

DIGITAL AGENTS

As hinted at in earlier sections on ubiquitous computing, the extreme form of this virtualization is to not only shift material travel and embodied interaction into mediated communication but to first automate and finally replace the worker with a digital software agent. If this is too pessimistic, one might speak of 'hiring-on' digital agents, but there is no doubt that *virtual workers* will compete with humans for work.

One current example is an animated character who performs the function of the newsreader (a predecessor was *Max Headroom*, an animated cartoon based on the idea of a computerized newsreader). Similar attempts have been made to create software-based characters who would function as models for fashion photography and digital fashion design. There is no requirement that such cyborg conform to a human standard. In fashioning 'Ananova', a digital newsreader, the programmers faced the problem of rendering hair – one of the most complex tasks because of its complexity and lack of algorithms by which the computer would display both the independence of single hairs out of place and the collective disposition of shocks of hair. Indeed it would appear that a number of these efforts are destined to fail because their ability to impersonate a human is never perfect. They remain locked within the virtual world of computer and graphical media. However, others surpass human qualities by adopting the model of the comic-book superhero. The popular Japanese anime character 'Sailor Moon' is one such model in which attempts to conform realistically to the variability of the human body are discarded in favour of cartoon qualities which allow something like hair to be merely suggested or to be coloured in any manner whatsoever to make a break with realism clear.

Others are animated composites, such as 'Webbie', an animated fashion model which may be used to produce both online and print-media fashion displays of clothing. Software such as 'Poser' now allows off-the-rack bodies and characters which the program can animate and display from any angle. While the continued popularity of celebrity models and international modelling stars suggests that humans are unlikely to disappear, one can imagine a role for the software-generated virtual models in giving clothing designers a three-dimensional and virtually living picture of the effect of changing fabrics or cuts. The ability to create animated scenarios of increasingly detailed and lifelike computer-generated characters raises moral and ethical questions when software is used to depict immoral and criminal acts or to show barbaric cruelty, sexual torture of children or the sadistic infliction of pain without endangering live people. Do non-human actors and virtual models or other digital agents demand ethical treatment if only because they represent and pose as humans? Does our unease at these possibilities for the misuse of the virtual hint at an underlying sense of *virtual ethics* for the workplace and for virtual entertainment spaces?

SUMMARY

Society faces major questions when it comes to virtual work. Virtual working involves new work environments – skills, times and places of work change in tandem with the types of work and members of the workforce. In virtualized work, tasks combine:

- *abstractly* coded elements, for example, symbols of computer screens; they involve
- *virtual* elements which may be neither tangible nor directly visible, only symbolized to the operator, while still involving:
- *material* performance of physical work by workers or by a machine at a distance.

Workers themselves are central to this process of the virtualization of work but neglected in many studies. Virtualized work is often experienced as imposing a distance, an alienation, from the actual world of material objects and face-to-face interactions with co-workers. The virtual figures in all forms of alienation, but the virtualization of work by technologies extends this process. Social relations are dehumanized as people struggle to keep up with the pace of machines and resort to more curt forms of salutation and command. Workers face demands on their lived relationships as well as on their bodies, and may suffer both stress and repetitive strain injuries (RSI).

Virtual spaces of labour are stratified and marked by material and embodied conflicts along the lines of gender, ethnicity, region, accent, professional status and pay. Lower-status technicians and support personnel may be invisible to users and higher-status professionals who may be equally frustrated. The possibility of ubiquitous computing further hides the operators and infrastructure of communication networks.

7

BUSINESS SENSE FOR A VIRTUAL WORLD

Chapter 6 considered the experience and roles of different workers in constructing virtual work environments and establishing and maintaining digital virtualities. These take on tangible form in the transformation of workplaces and in the spread of teleworking. This chapter concerns the global scale of a virtual economy and its manifestations in specific organizational arrangements. Beginning at the global scale, it focuses down to the level of firms and organizations before presenting general lessons on the relationship between online virtual services, traditional virtualities, and actual success and profitability of firms and organizations. It proceeds as follows:

- Virtualism in economic thinking.
- The economics of digital virtualism.
- Digital virtualism and financial institutions.
- Information and attention deficits.
- Online rumours.
- The experience of companies.
- Managing the virtual and the actual.

ECONOMIC VIRTUALISM

Economists, financial institutions and their customers have long been familiar with a product that is virtual: money. Monetary values fluctuate against each other and are subject to a variable rate of conversion into actual commodities, depending on inflation or deflation. An example is the variation in the cost of wheat which may reflect supply and demand but also reflects the conversion rates between one's currency and that of the producers of the commodity in everyday life. Payments are made using pieces of plastic, balances are read on computer screens or on bank statements, and the value of a piece of currency, whatever its denomination, is not the actual cost of printing it, but what it can be exchanged for – a virtual value (Simmel, 1990).

Some economic theorists have pointed to the increasing degree to which the economy is understood in a disembedded manner. Disembedding means that economic activities are treated as disconnected from all other spheres of activity. The cultural and social aspects of work, for example, are neglected in the cultural and historical attitudes of advanced capitalist societies. Purist neoclassical economists is an extreme case (Stewart, 1995); the media, however, are also dominated by a view which treats interest rates, unemployment and productivity without reference to their embeddedness in particular societies and localities (Callon, 1998).

Disembedding is a form of idealization. Although they conflate the abstract with the virtual and do not examine the further implications of the shift, Carrier and Miller agree that the situation of treating abstract economic theories as parsimonious and actual is a form of economic '*virtualism*' (Carrier, 1998). Referring to IMF debt restructuring policies (see also George and Sabelli, 1994), Miller comments that economics is now the ultimate political authority disguised in a disengaged, abstract rhetoric:

> While capitalism engages with the world and is thus subject to the transformations of context, economics remains disengaged

> ... because economics has the authority to transform the world into its own image. Where the existing world does not conform to the academic model, the onus is not on changing the model, testing it against the world, but on changing the world, testing us against the model. The very power of this new form of abstraction is that it can indeed act to eliminate the particularities of the world.
>
> (Miller, 1998: 196)

By maintaining abstract models of rational economic actors and of consumer choices, an abstract model of the economy is constructed and then inappropriately treated as virtual; that is, as accurately representing the actual processes of markets. Miller concludes that since consumption is the main opportunity for most individuals and families to benefit from contemporary forms of capitalism, 'the move to greater abstraction had to supplant consumption as human practice with an abstract version of the consumer. The result is the creation of the virtual consumer in economic theory, a chimera, the constituent parts of which are utterly daft' (Fine, 1995; Miller, 1998: 200). Although this confuses abstraction and virtualization, the thrust of the argument is accurate. An army of highly paid management consultants implements the virtual models of economic activity and the virtual consumer on to actual market and business situations, often negating a firm's staff's knowledge of concrete conditions in local markets – with disastrous results for the business (Miller, 2000).

Indeed, neoclassical economists make no claim to represent flesh-and-blood consumers but only aggregates of consumer behaviour. 'Their protestations of innocence are hollow, however, because these virtual consumers and the models they inhabit and that animate them are the same models that are used to justify forcing actual consumers to behave like their virtual counterparts' (Miller, 1998: 200). The result is the self-alienation of a major area of human endeavour from questions of human welfare.

The economics of digital virtualism

The virtual organization is the name given to any organization which is continually evolving, redefining and reinventing itself for practical business purposes (Hale and Whitlam, 1997: 3). There is no agreement as to what binds together such concepts as virtual organization, offices, corporations and factories (Jackson, 1999: 10), except that they

> are associated with the use of cybertechnologies to allow people separated by time and distance to work together cohesively. The concept of virtual organization is therefore encapsulated in a desire to use information technology to enable a relaxation of the traditional physical constraints upon organizational formation and adaptation.
>
> (Bartnatt, 1995: 4, quoted in Jackson, 1999: 10)

The explosive growth in information technology spending drove a late twentieth-century economic boom adding a full percentage point to annual US economic growth in real terms since 1994 while lowering inflation at the same time (estimates put this at about half a percentage point a year (Rubin, 2001, Canadian Imperial Bank of Commerce). The 'bust' that followed in the USA and its closest trading partners mirrored on a less drastic scale the similar implosion of the East Asian economic bubble, a crisis attributed by some to the terms of support and loans offered by the IMF. In countries which had been labelled the 'Newly Industrialized Economies' (NICs) such as Thailand and South Korea, telecommunications infrastructure and the export of IT components had played a similarly important role in propelling growth, and had even allowed the emergence of a middle class in these countries, changing their politics forever. The diffusion of information technology in businesses was estimated to have raised the annual growth in productivity by two-thirds of a percentage point over the 1990s (US Department of Commerce estimates).

Although the 1970s saw the invention of the microprocessor and the 1980s saw the adoption of desktop computers in the businesses of Western, developed nations, it was the 1990s when IT spending had an economy-wide impact and a cultural impact. Personal computers (PCs) became common in households, and portable computers, email devices and mobile phones were widely adopted if at varying rates (Figure 7.1). For example, by the early 1990s most Canadian homes (90 per cent or more) had a modem-equipped PC, while in Finland most people (90 per cent+) owned a mobile phone which included a database of personal phone contacts and could handle email, text messaging, and often had an electronic agenda. I have argued in preceding chapters that the cultural impact of the spread of IT is a renaissance of virtuality supported by digital information and telecommunication. Virtuality is not dependent on computers per se. It could equally well be enabled by mobile phones or even

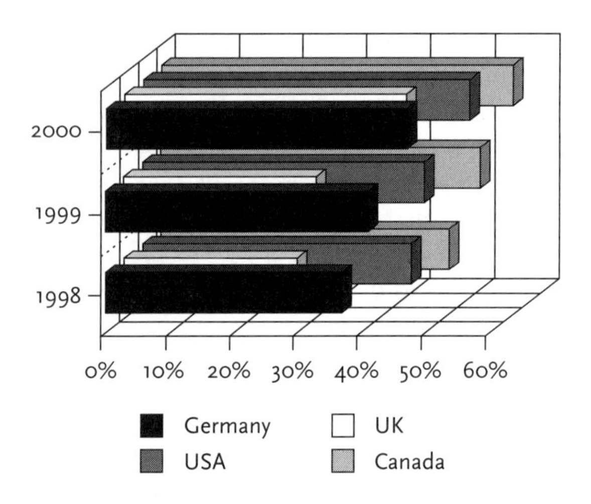

Figure 7.1 Home computer ownership 1998–2000
Sources: Federal Statistical office Germany, 2000 (www.destatis.de)
National Statistics, 2000–01 (www.statistics.gov.uk)
US Census Bureau, Public Information Office (www.census.gov)
Statistics Canada, Income Statistics Division, December 11, 2001 (www.statcan.ca)

via shorthand codes sent to pagers as was common in Japan and Korea in the early 1990s (Shields, 1997). The economic impact is part of the cultural shift to accept virtualities as concrete objects and events.

DIGITAL VIRTUALITY: BANKS AND BROKERAGES

It is striking that there have not been greater strides in the digital transformation of money. The creation of a system of digital virtuality would seem tailor-made for the management of a virtual object which is already dealt with electronically. Furthermore, the Internet affords greater efficiencies for lenders who might deal directly with borrowers rather than through costly intermediaries such as banks. Automated banking or teller machines (ATMs) have already accomplished this on behalf of banks themselves, bypassing actual human beings in transactions with customers. Telephone- and web-based facilities are becoming more common despite fears over the security of web-based transactions.

However, except for share-trading, Internet-based financial retailing scarcely exists. Back-office transactions are still accomplished on in-house electronic systems and trading is done on closed, proprietary systems, such as the City of London's electronic trading floor, launched, with much fanfare, as 'the Big Bang'. In part this is due to the power of banks to deny newcomers entry to the field and to suppress new approaches (cf. resistance to online payment systems such as PayPal in the US (Anon., 2002)). Banks and their staff have been unwilling to change old ways of doing things that have generally been extremely profitable and robust. The conservatism of banking has been challenged by a new breed of entrepreneur and by firms which have pioneered online stock trading, such as E*Trade and Charles Schwab. For example, Merrill Lynch

> held back from introducing an electronic product, mainly because of the 'channel conflict' between electronic distribution

> and the 17,000 retail brokers who constituted the core of Merrill's competitive effort. This army generated billions of dollars in commissions. But that was only a fraction of their true worth. As part of the biggest single distribution system for securities in America, they made Merrill a favoured participant in lucrative corporate and municipal underwritings.
>
> (*Economist*, 2001b: 79)

However, the older, 'stuffy' banks and securities firms can pride themselves that even after the wholesale evacuation of Wall Street in the immediate aftermath of the terrorist attacks on the World Trade Center and the accounting and investment firms located there, the system continued to operate. Even the most well-heeled and sophisticated online trading systems have had problems: on 5 April 2000, the day after Microsoft had first been found guilty of anti-competitive practices, and the last day of the fiscal year for investors, the London Stock Exchange computers went down for nearly eight hours. Every day, users complain that technologies are too slow, inconvenient and require multiple tries to accomplish simple tasks. By contrast:

> 'Legacy systems' . . . have tended, by and large, to work. Big banks process trillions of dollars a day. It is almost inconceivable that they might close down for a few hours because some clever internet saboteur has found a way of snarling up their technology (as has recently happened to some of the biggest websites).
>
> (*Economist*, 2000b: 6)

Yet the opportunities to reduce costs via the efficiencies of web-based transactions, where customers can be induced to do their own form-filling and transacting, promises greater profits. Once their development is invested in, customers will be bribed and bullied online (*Economist*, 2000b). One officer with Charles Schwab explains its success based on demographics:

> 'the baby-boomer generation . . . is unwilling to compromise and needs to feel "empowered" and in control . . . a baby-

> boomer confronted with an easy-to-use Apple computer is not lost in admiration for the clever people who designed and built it. He thinks: "I am a computing genius!" Online broking can turn them into investment geniuses as well.'
>
> (D. Leemon, quoted in *Economist*, 2000b: 15)

However, the new online discount brokerages have not been particularly profitable, many only breaking even at the height of the bull market in 2000. After the technology boom and the novelty of online trading, unsustainable volumes (Credit Suisse First Boston, cited in *Economist*, 2000b: 15) fell after the collapse of the dot.com market 'bubble' and the bear market that followed for several years. In the UK and Germany, widespread share-ownership was a new market which took off at the same time as Internet use, whereas in North America, online trading was led by investors moving from other channels such as discount brokers (G. Seidel for Arthur D. Little, Germany, cited in *Economist*, 2000b: 16).

Online trading is not only a question of financial transactions via personal computer: almost every company offering online financial services also services another type of interface, whether phone or palm computer (or personal digital assistant (PDA)). Not only are wireless devices supported but access via gaming consoles such as Sony's 'PlayStation' and interactive television is possible, depending on cultural preferences and the market. In Finland and some other European and Asian countries such as Japan, banking and broking is offered on the small screens of wireless application protocol (WAP) and third-generation (G3) mobile phones. WAP devices support simple menus rather than complex webpages. However, in 2001 Japan still lagged behind South Korea and Taiwan in the use of online brokerages, boosted by the rebound after the IMF crisis.

In North America, PDAs such as the Palm Pilot are favoured, offering slightly larger but still only notepad-sized screens which are more than adequate for the display of columns of numbers. The devices of 'ubiquitous computing' may take many forms but the common vision is the possibility of accessing information at any time and any place. The flip-side is to make users accessible to messages, advertisements and reminders at all times; a telemarketer's dream. For example, out of many of its failed Internet projects, Merrill Lynch, the stockbroking firm, created a broadcasting studio to create web-based video programming featuring comments by analysts and interviews with bosses of firms offering stock placements and so on (*Economist*, 2001b: 80). The success of these sites, despite the contraction of the market in 2001 and 2002 which dampened interest in popular investing services for small investors, has indicated the direction in which multi-channel approaches a future (likely more upmarket) clientele (*Economist*, 2002: online). *The Economist* offers the following suggestion of how it might develop further:

> Suppose you are a potential investor in a company's initial public offering of shares, and have just finished watching the boss boosting his company's prospects on Merrill Lynch's online investment banking service. The phone rings. And yes, it is a Merrill Lynch salesman who knows you have been watching, and thinks that now may be the moment to clinch a sale.
>
> (*Economist*, 2000b: 9)

With the huge pressure on attention that the stock-market represents, and the marketing and cross-selling efforts of the computer-guided sales reps and call centres of financial retailers, a challenge is thus posed: how to manage the place of the digitally virtual economy in our actual, day-to-day lives.

VIRTUAL INFORMATION: ONLINE RUMOURS

News services such as Bloomberg and Reuters have been complemented by online advice, financial data and gossip in chat channels and Usenet newsgroups. In the welter of information, it becomes increasingly difficult to know which are the reliable sources and to distinguish relevant information from deceit, fraud and trivia. Despite the traceable character of many online messages, the democratization of investment and the spread of day-trading has allowed old-time cons to move online. Many have attempted to become professional day-traders, but most are computer-literate white-collar workers. For these workers, the stock-market has taken pride of place over computer solitaire or simply a novel with which to waste company time. Workers can take advantage of stock-price tracking programs which post an up-to-the-minute chart of price fluctuations of one or a few specific stocks, alerting the day-trader to short-lived fluctuations which can be the basis of small, daily profits.

Virtual information is 'almost correct' – in other words, *gossip*. What is said on Internet chat forums, where people post messages about stocks, impacts on value quickly: like the panic which ensued after Orson Welles' 1938 radio drama *War of the Worlds* that took the form of a newscast of an invasion from outer space. Widely held stock is less likely to be affected, but other stocks have seen 30 per cent increases on speculation that a chat-room posting is true (e.g. the case of Pairgain on 7 April 1999), especially if the rumour is picked up by the press. For example, the US Securities and Exchange Commission urged investors to 'take what they see over the chat rooms not with a grain of salt, but with a rock of salt' (A. Levitt Chairman, SEC, cited in Karleff, 1999) as speculators drive stock up or down for their own benefit.

It should not be surprising in a business built on trust that people are willing to pay for human service. Other people, while not guaranteed to be reliable, are still the most compelling bearers and 'interfaces' of information. Thus online stock-trading systems built for the affluent and offered by brokerages such as Merrill Lynch have faired poorly. 'In retrospect . . . the Internet as an investment . . . turned out to be a disaster, and the Internet as a means of making an investment . . . could be more of an opportunity' (*Economist*, 2001b: 79).

DIGITAL VIRTUALITY AT ACTUAL FIRMS

Companies large and small tend to under-manage the relationship and flows between the concrete and virtual in their enthusiasm for the virtual, which gets over-hyped and over-managed. Virtual business includes challenges to organizational cultures. For example, a firm's managers, and outside analysts, described the difficulties of converting to e-business models including use of email rather than paper memos and web-based procurement and sales:

> Managers had to watch carefully for reprobate employees using 'parallel paths' (the telephone, for instance, or a walk to a store) to order supplies, say or arrange travel. Some offices even closed their mail rooms for all but one day a week (and that only for the incorrigible legal department) to stop employees from using regular post. Others locked their printer rooms except for occasional days when bosses would station themselves at the door and demand from those who came through an explanation for their sad inability to shake old paper habits.
>
> (*Economist*, 2001d: 76)

Yet it is inefficient to force all material procurements through virtual channels. Enforcing the virtualization of business relationships without regard to issues of trust is a further problem. The resistance offered by internal organizational cultures is an important sign that managers have not typically been trained to notice. Preferring computer-mediated communications over telephone is similarly not necessarily an aid to virtual business, because it is overly focused on supporting the virtual via one technology rather than drawing on all that are available and appropriate. Organizational cultures are already forms of *virtual infrastructure* which facilitate communication, problem identification and solving by allowing transactions and exchanges to proceed without having to reinvent protocols, etiquette and language again and again.

General Electric

For example, electronic commerce existed from the 1980s in the form of proprietary business-to-business ordering systems such as General Electric's Information Services (GEIS) which, even after many had shifted to Internet-based systems, accounted for over US$20 billion annually in transactions (*Economist*, 2001d: 75). The system is a centralized messaging system more similar to highly structured email than the direct business-to-business (B2B) Internet model (about which more below). Such systems provide *protocols* which structure the opportunities for greater efficiency in digital–virtual rather than analogue or face-to-face transactions. Standardized protocols need to be seen as an important form of discipline for digital–virtual systems. They accomplish this by yoking the virtual to standardized material requirements and needs. At the same time they build in or freeze the status quo with all of its inequalities and unevenness in terms of access.

As part of GE's large conglomerate, change was difficult at GEIS: 'Its data centre in Rockville, Maryland boasted a NASA-like command centre run by former military technicians still sporting crew cuts and smartly addressing supervisors as "sir"' (*Economist*, 2001d: 75). However, this old-style mainframe computer system securely served 100,000 companies and collectively processed more than a billion transactions annually. GE's network, the Electronic Data Interchange (EDI), offered automated machine-to-machine transactions for large clients such as Wal-Mart, one of the largest discount retailing chains. The model is described as follows:

> EDI saves companies money, mostly by taking people and error out of the equation, but it does not fundamentally change the way business is done or who it is done with. That changed when the Internet came along. B2B exchanges did more than simply take existing relationships and transactions and turn them into digital form: they offered the potential for new relationships and

> new sorts of transactions, from auctions to direct sales without middlemen and brokers.
>
> (*Economist*, 2001d: 76)

A web-based system would allow this service to be offered to smaller firms and to many of GE's customers, and to compete with the over 700 web-based B2B exchanges that were founded in the five years before 2000. These often operated on an auction basis, much as does 'E-Bay', allowing sellers to find hitherto unknown buyers. Still, 'Buying from suppliers online was one thing, but selling to customers online risked putting GE's sales force out of business. Big Internet investments risked hurting each division's bottom line, against which bosses are mainly measured. . . . The greatest hurdle has been not technology but culture' (*Economist*, 2001d: 76). Obviously the Web would impact on the human sales force, moving some into customer call centres, perhaps 'downsizing' the overall size of the sales force, and changing the way remaining sales staff operated. The sales staff were resistant because they understood the relationship between the virtualization of buying and selling, and the material world of personal visits, paper contracts and regionally responsible sales reps who might even follow up with customers upon delivery.

The flip-side is that there is a danger in using IT as a way of trying to do more cheaply the same old things in the same old ways, just by virtualizing them – but things are never 'just the same'. New management and corporate governance strategies are required to respond to the split-second pace of digital virtualities, and the nuance and symbolic basis of the virtual. One cannot take an old 'hands-on' approach when one is dealing with intangibles such as information.

There are also implications for the firm itself. A prime reason why economic activity is organized within a firm rather than in an open market is the cost of transactions and communication. If the Internet cuts the cost of communication, then organizations ought to be able to do less in-house and to outsource and subcontract more. 'The internet may make a conglomerate cheaper to run; but it could also make other forms of organisation cheaper still' (*Economist*, 2001c).

The collapse of Enron

Understanding the virtual and its relationship to the concrete is essential in today's economics. The key term that surfaces in much of the economic literature is the notion of *intangible assets* and markets that are composed more of information and ideas, promises of probable returns in the future, and even completely abstract speculation, than concrete, fixed assets. The challenge involved is that virtual assets and values (e.g. futures options on a commodity such as oil at a certain price) must be continually and smoothly converted into real assets (in this case oil materially delivered, in the grade and quantity required, when and where it is needed). Promissory notes must also be converted from and into cash. There is always a risk of a 'run on the bank', as happened in the case of Enron, a corporation which was both an energy commodity market administrator and a partner in trades. Beginning from its roots as a wholesaler of energy commodities, Enron provided an online trading system with itself as the middleman in each trade. It had to acquire sufficient material and financial assets to ensure liquidity and be able to guarantee the fulfilment of trades and contracts. The essence of the system was that Enron acted as a guarantor on the basis of buyers' and sellers' faith in its size and integrity.

'Enron itself . . . provides the market's liquidity, and the firm's good name ensures certainty of contract execution' explained *The Economist* which celebrated Enron's success in an '*E-Strategy brief*'. While the bosses misunderstood this as merely a reinforcement of the business model of facilitating energy trades, others argued that as Enron branched out into paper pulp, liquified natural gas, crude oil and coal trading, 'With each new trade, it has less and less to do with energy, and more and more to do with making markets . . . the firm's goal is "the commoditization of everything'" (*Economist*, 2001a: 72). The increasingly virtual character of the company and its attendant risks were not understood.

Profits appeared inflated by counting each transaction in the trading system as income rather than only the fee that Enron earned on the transaction. At the same time affiliated companies were created whose liabilities were thereby kept off Enron's balance sheet because of a loophole in American accounting procedures, even though Enron was ultimately liable. Analysts argue that this was only compounded by the auditing of Enron's accounts. Auditors and management consultants had become important players in arguing for the allocation of resources within firms by generating new 'narratives' of corporate culture – virtual objects if ever there were (see previous chapters; also Power, 1994; Salaman, 1997). In an ironic twist, the growth of auditing itself was legitimated by reference to the welfare of consumers (MacLennan, 1997). These consumers are the virtual citizens of 'consumer democracies'. Thus, Miller argues, 'policies justified in the name of the consumer citizen become the means to prevent the consumer from becoming a citizen, from determining the priorities of expenditure in the public domain' (Miller, 1998: 204).

When questions were raised about its accounting practices, both investors and banks that had backed the company withdrew their investment, leaving Enron suddenly bankrupt in one of the biggest business scandals ever seen. Each trade of a commodity involved not only the virtual but the assumption that it was backed by an existing concrete quantity of commodity, with elements of the probable included in the form of calculations of potential price spreads and trends (buying low now to provide capital for companies that would move goods and deliver them later, to be sold on at a higher price). Accounting practices, however, allowed the purely abstract and non-existing to enter the books as Enron gambled that not all actually-existing (concrete) assets would be required at any one time.

What was offered as virtual (real but ideal or representational credits and account balances) proved to be untrue and non-existing (that is, what we called in Chapter 1 a non-real representation, a figment). When the trust that virtual assets – promised deliveries, probable prices, the online environment itself – could be converted into concrete commodities or into the similarly virtual form of money was challenged, the system collapsed.

The case of Enron reveals that companies operating in the virtual not only face challenges to their organizational culture but also to their management systems. Yet Enron is a good example of the rush to turn all business processes into digital services which typified the late 1990s. Every company was advised to turn itself into an 'e-business' based on offering its existing strengths as online 'e-services' and to build up an 'ecosystem of partners'. However, auditing and accounting practices failed to keep up with the complexity of the partnerships and alliances that resulted.

MANAGING THE VIRTUAL AND ACTUAL

Like the financial sector, companies in the late 1990s could be divided between the dot.com upstarts and the established 'bricks and mortar' businesses. The Internet hype of the time and the flight of investment to the dot.com bubble encouraged every business to get online and virtual as quickly as possible, leading to the problems of neglecting the material bases of the enterprise. The mantra was that the Web displaced traditional bricks and mortar brands. Established firms such as Merrill Lynch and Charles Schwab (above) hastened to get online. The mantra was 'clicks for bricks'. Now, after a shake-out of dot.coms without products or profits, firms are thinking about permanence – 'bricks for clicks' – and the bottom line. Mergers such as that between AOL and Time-Warner were designed to expand customers and marry the virtual and concrete strengths of the respective partners.

Established businesses have begun to think beyond promotional websites to a new environment in which contact with clients and customers is ongoing, via many different types of devices and media. The new 'clicks and bricks' paradigm was first described merely as a matter of 'convergence'. This buzzword captured the sense of a middle ground as well as the convergence of computing with entertainment, the Internet with broadcasting and so on. With hindsight, however, one can see that convergence is a question of a new type of company. Rather than worrying about 'channel conflict' as in the case of the full service brokerages (above), the argument being made is that the most successful model is focused on how a firm relates to its individual clients. Earle and Keen agree, giving the example of the car:

> Automotive firms are beginning to understand that the car – the product – is a platform for delivering services. By delivering customer relationship services such as in-car navigation emergency roadside assistance, or telecommunications, car companies can deepen their bond with customers and also generate more revenue. . . . It's in the combination of the product and the service that revenue and profit are being made. It's not that the product is any less important; it's just much more useful with a service wrapped around it. Services drive the customer experience.
>
> (Earle and Keen, 2000: xi)

Whereas, for example, an online bookseller like Amazon.com began with relatively static websites presenting information and allowing ordering, these were made into an interactive social world where buyers and authors were encouraged to add opinions and reviews. Instead of a storefront or broadsheet approach, webpages are now more like mosaics of customized information, assembled dynamically to offer each browser a set of customized offerings. They are less like print ads and more like electronic travelling sales reps at the door.

The virtuality of the digital-driven aspects of all organizations is beginning to transcend its technological ghetto of IT departments and to be seen as a key part of the traditional, human, virtual aspects of an organization. Traditional virtualities include trust and goodwill, brands, and organizational culture. This brings in the full range of management and skills, and allows the digitally virtual to be understood in relation to the material activities of any enterprise, whether public or private.

BRANDS AND RELATIONSHIPS

Branding, online and off, provides a good example of virtuality. Where the product itself is a material object or perhaps a service with concrete results, brand names are virtual. As discussed in earlier chapters, the virtual is both real and ideal: it does not have the tangibility of the actual but is something that clearly exists none the less. Customers are not purchasing an abstraction (such as a promise; that is, a non-existing ideal) but an *intangible* good. Brands offer a form of guarantee against risk. The promise of consistency (such as knowing the menu and decor will be the same quality and type at every McDonald's) is indexically summarized in standardized corporate logos. Brands and their logos extend commodity fetishization by carrying the full range of meanings that might be associated with a commodity into a symbol. The life of brands is intimately related to the media and other channels by which reputation is spread, such as gossip. The question of reputation, however, is one that must be maintained either by the satisfactory performance of goods or by disinformation in the media to persuade unhappy customers that an unsatisfactory experience was not the fault of the brand or its parent corporation. Whereas one might use up a product, a brand remains in the mind long after the commodity has been discarded from everyday use.

Writers such as Michel Maffesoli (1996) have pointed to the use of brands to signal the values held by individuals and by groups. Brands may be the basis of affiliation (such as football (i.e. soccer) fans who fetishize Armani jackets or those who favour certain marques of sports clothing) with others, and with the characters and values portrayed in the media. One 'brands' oneself, as it were, marking oneself as a particular type (Shields, 1993).

Successful online business models focus on relationships rather than on a single transaction. Rather than concentrate on price-cutting to conclude a single transaction, relationships and collaborations and added value to clients are the key to repeat business. While this is a risky strategy which requires superb delivery and follow-up, they are less vulnerable than the low-cost oriented transaction model to competitors who have offered services completely free online in order to build up relationships.

The case of online portals such as Yahoo! or AOL is a good example. They challenge existing brands by subsuming them as one aspect of an overall service relationship. They became the best-known online business successes based on their model of charging advertisers while offering the use of search engines for free. This has expanded to free websites, email addresses and a 'My Yahoo!', a customized homepage for one's browser featuring updates on information you had selected. Web-based collaboration tools such as 'chat rooms' and access to shared files has continued the logic of building up an ongoing service relationship in which established brand names may be subsumed under the portal's name.

Earle and Keen (2000: 17–19) argue that the relationship and branding models are complemented by a number of other key factors or value drivers at which online services excel. Logistics such as inventory management and procurement is exemplified by the case of GE's online ordering system. Brokers or intermediaries are the basic model of a stock exchange, which also appears in the case of Enron. They offer new channels which complement existing ways of accessing information or services. Finally, financial dynamics may be changed to transform costs and margins, and the ways in which prices are set – for example, allowing customers to bid for goods rather than stipulating prices as in the online operations of Priceline or eBay. In short a set of new approaches, contrasting with established practices (Table 7.1).

Table 7.1 New approaches, contrasted with established practices

New practices	*Old practices*
• cultivate long-term customer relationships	• focus on transactions, cutting prices for single sales without reference to service
• become a value-adding intermediary between sellers and buyers	• position within industry, production
• perfect logistics and integrate supply chain	• administration based on functional areas
• build a brand to differentiate and position services	• brand equity
• harmonize all channels on behalf of the customer	• restricting access, Web as only channel
• transform finances so that maximum revenue is generated on minimum capital and sunk costs	• expense management • price/earnings ratios

Despite the cynicism of the twenty-first century, these remain important imperatives shaping the strategies of all businesses as they struggle to manage both the concrete and the virtual. 'The com mentality made revenue growth the target – build traffic and sales. IPO fever and Internet hype talked as if economics were irrelevant' (Earle and Keen, 2000: 143). Rather than measure success based on profit and loss or return on capital, dot.coms assumed that increased market share was the basis of profitability. This is the reason why new start-ups spent heavily on advertising – 65 per cent of revenues for an online retailer compared to 5 per cent for a traditional bricks and mortar merchant (Boston Consulting Group, cited in Keen, 1999) – and were much criticized for this later on. The 'burn rate' of funds raised through initial public offerings (IPO) was not only for the technology but also for marketing: $1.5 to $3 million to set up an interactive sales site, $15 million for a financial services site (Gartner Group, cited in Halper, 1998), $5 million a year to launch a portal, $300 million over several years to build a brand name (Kahn, 1999, cited in Earle and Keen, 2000: 145).

The defence offered by those involved in successful online enterprises is that it is hard to make sense of start-up companies because they are so new. Their marketing should be seen as an investment in building a consumer awareness and relationship, not a cost. Heavy initial investment in R&D may build intellectual capital but counts again revenues, dragging down profits. 'Just as Intel, for example, could more than double its earnings immediately just by halving its R&D' but would lose its market and its technological dominance would suffer the following year (and its stock would slump immediately), it is short-sighted, they argue, to demand that an online company such as Amazon become profitable by cutting its marketing (Earle and Keen, 2000: 148). In many ways the start-ups have a huge tax advantage, and, whereas advertising a website is immediately deductible, putting up a new store is a cost which is depreciated over many years.

Unfortunately this is a speculators' model. In the inevitable shake-out, many firms collapsed, and only a few were left to consolidate their online services. For these 'gorillas', the logic worked out as *Red Herring* explained:

> The biggest benefit of being a gorilla in your category is the ability to access low-cost capital either through private investment or the public markets. This position also provides Web entrepreneurs with high market capitalizations that they can leverage to acquire new services, along with privileged partnership opportunities that can expand their audience reach.
>
> (Perkins, 1999: 14)

Firms ran out of funding when investors refused to put more cash in without signs of profitability. Many firms went broke before becoming profitable, leading to a business model which focused on positioning the company to be acquired by the dominant and best-funded 'gorillas' in their sector.

In the established model, capital is the main asset, but in the new model, capital is seen as something that has a high carrying cost. Instead, a positive cash flow (collect from clients now, pay creditors and suppliers a month later) replaces the role of capital. The effect is to deliver profits from an investment far below what might be expected – converting virtual potential into material returns. Yet the result is that initial investment costs will mean that new online companies will record low and negative profits, and no timeline exists apart from investor impatience. Hence the high losses of popular successes on the Internet – Amazon.com, Yahoo! and others. Amazon.com reported that net sales for 2001 fiscal year reached a record-setting $3.12 billion, a 13 per cent increase. Yet the fiscal 2001 pro forma operating loss was $45 million which the firm counts as an improvement of more than $270 million over the previous year's loss (Amazon.com, 2002). This defence however unravels as firms such as Amazon have grown bigger outstripping the scale and experience of traditional book distributors. In order to deliver – to transform their virtual advantages into material transactions – they are forced to build warehouses. Amazon is reported to spend an estimated $300 million to build distribution super-centres, increasing its invested capital and forcing it on to a valuation path more similar to established corporations.

We now know that the new economy is not dot.com versus bricks and mortar but a synthesis of bricks and clicks:

> dot coms spurred traditional companies to make use of the internet to digitise their business processes and become more innovative. . . . At the same time, dot coms started adding bricks to their clicks so as to honour their customer promises. As companies became more virtual, dot coms became less so.
>
> (Kim, 2001: 10)

Pressures on online companies confirm that digitally based, virtual goods and services must be managed in relation to their material counterparts to ensure actual profits. This translates into key highlights for managers (see also *Economist*, 2000a: 38–39). In managing the virtual and actual together:

- Ethics matter: the relationship between the virtual and concrete involves ethical duties.
- Customer-focused relationships involve ongoing services accompanying products and integration of services into everyday life.
- Online virtual services must be understood in relation to traditional virtualities in provider–client relationships such as trust, courtesy, service and perceived added value including thrift and exclusiveness.
- Collaboration at all levels appears to be a key feature of the business landscape: celebrity teams (discussed in previous chapters), alliances and partnerships enabled by online information exchanges and actual exchange of workers.
- Speed and contraction of design to market and product cycle times are factors.
- Intellectual property and the value of knowledge accessible inside and to the organization is a key asset, but clarity of content and context remains an issue.
- Protocols represent a new structure to opportunities and a form of discipline to channel information.

SUMMARY

This chapter has critiqued the virtualism of economics and current management thinking. It reveals a misunderstanding of the importance of keeping a close relationship between the virtual and the concrete aspects of organizational efforts. Economic Virtualism occurs when economic theory abstracts from capitalism and markets, but this idealization has been treated as actual – that is, as a set of virtual laws to be imposed on actual firms and economies. Rather than either/or, the name of the game is managing a relationship between the virtual and the concrete for advantage. Just as previous chapters noted the strategic importance of low-status workers at points of translation from one network to another or from the virtual to the actual, so the transition back and forth from the material world of products, services and transactions to the virtual and abstract world of orders, bids and client information are key points of risk, exemplified in the example of Enron, for organizations and businesses.

The conclusion of this chapter may be that the digital–virtual has been, overall, a negative development for most businesses. Banks and brokerages have had mixed results in taking advantage of the virtual qualities of money to create digital–virtual methods of investing and managing money. Online transactions remain relatively difficult with many concerns about security. While some enterprises come into existence on the basis of new opportunities, more firms are plagued by problems of speed and pace, an increased loss of control of information about their activities through online rumours, and problems of management misunderstanding and error. In attempting to impose digital virtual forms on more appropriate but face-to-face forms of actual, material–world interaction and on traditional virtualities such as corporate culture guiding narratives, motivation and self-understanding can easily be damaged.

8

RISK CULTURE, TRUST AND THE VIRTUAL

This chapter argues that the virtual plays a central role in assessments and decisions concerning risk in everyday life. We will consider the notion of a risk society, and of increased anxiety in Western societies in terms of abstract, virtual, material and probable threats. I argue that the virtual figures in risk when threat is understood to be widely present, as in the case of ecological and health risks due to pollution. In what follows we examine the elements of virtuality to our calculation of risk, and our sense of vulnerability or security. This argument involves the following steps:

- Risk is virtual as much as actual (i.e. concrete and/or probable).
- The notion that we are living in a 'risk society', permeated with an awareness of threat and an obsession with security, suggests that a 'risk culture' is now the context in which risk is understood in everyday life.
- Media play a key role in the circulation of information.

- Risk is usually understood in terms of probability and concrete danger, but decisions about risk include components which are virtual (trust), and abstract (for example, the security industry and pursuit of 'international security' in general).
- Insecurity is a virtual overlay on the projects and routines of everyday life.

Risk is always more than concrete danger and calculations of probability because of the importance of perception and understanding as ingredients in risk assessment. The virtual and abstract are drawn into the understanding, evaluation and reaction to risk as a social construction. Thus some may be more risk-averse, and others actively court risk in the form of danger as a source of excitement.

> In complex situations, the calculation of the agent *might* have little to do with the real [concrete] risks of the situation but be nevertheless decisive for the agent's decision. Risks can be overstated or underestimated. Risk assessments should be analysed as social constructions which result from the actor's understanding of the situation which is the result of interpretation and principally open to beliefs and manipulation.
>
> (Beckert, 1999: 22)

Our sense of risk is constitutive of the choices made in everyday life, and is a key force affecting the whole economy and culture of societies. In the United Kingdom in 1988, for example, the health minister claimed that eggs were infected with salmonella with the result that consumer purchases of eggs fell by half (BBC, 1998). The 2001 terrorist hijackings of aircraft in the US on 11 September had such a negative impact on travel – particularly on business and work-related trips by plane – that economic support for the entire industry was required. Because of the complexity of technological systems – from elevators, to city water purification and supply systems, to the computerized management of air traffic – individuals not only feel that matters are out of their control, but any one person lacks sufficient information to make knowledgeable calculations of risk. All people necessarily fall back on hunches, 'gut feeling', partial information and comparison with others' actions. One leaps to a conclusion, and in such situations mass hysteria and over-estimation of danger are common. We ask, 'Why take the risk?' And, although we may relish our encounter with some risks in extreme sports or in gambling, when it comes to their families people generally become much more cautious – 'Why subject my children to this risk?'

Political struggles involve representing risks in ways that they become palpable dangers which anyone would be foolhardy to tangle with – for example, requiring gruesome pictures of diseased body parts on cigarette packages. Despite the presence of sophisticated attempts to quantify risk in order to develop insurance markets to indemnify against it, the social management or calculated avoidance of risk are closely linked to the development of a sense of security, and of trust in social institutions and technological systems. When these themselves become threats – as in the case of nuclear power, or ozone depletion and subsequent skin cancers, or covertly-funded insurgents who return to haunt the powers that be as well-schooled terrorists – societies may be described as 'risk societies'.

RISK AVOIDANCE AND 'RISK SOCIETY'

This section provides a brief introduction to the 'risk society' thesis. This hypothesis proposes that a generalized change in the importance of risk and social attitudes towards risk marks a shift away from an earlier and more optimistic epoch of modernity. Ulrich Beck's diagnosis of a second phase of modernity or 're-modernization' (Latour, 2002) has been vulgarized in translation as a description of 'risk society' (Beck, 1992). 'Risk society' is a condition of social change driven by the unintended side-effects of industrial modernity (or 'first modernization' in Beck's term (Beck, 1992)).

> According to Beck, then, risk management is a defining characteristic of our age, realised in all domains from family and interpersonal relationships and employment insecurities, to environmental hazards and scientific practices. Risk Society is dominated by a narrative of the dark side of modernisation and the Enlightenment and, in particular, the constitutive role of science and knowledge within it.
>
> (Banks *et al.*, 2000: online)

After the Industrial Revolution in Europe, the expansion of production and the distribution of wealth and resources in societies came to be seen as the basis on which a just society could be achieved. 'Redistributive justice' (Rawls, 1993) aimed to achieve a more equitable distribution of wealth across socioeconomic classes. Visions of the 'good life' stressed hard work and rewards in the form of consumer goods bought on the retail market (Shields, 1993). Marx articulates the politics of nineteenth-century, European first modernity. He stressed the struggle of the conditions of production organized paradigmatically around divisions between workers with labour to sell for a wage and a bourgeoisie which controlled the means of production, such as financial capital and machinery. However, 'Class and other systemic differences within modernity . . . give way . . . to differences of knowledge within the information order' (Tulloch and Lupton, 2001: 11).

As negative environmental, medical and dehumanizing impacts of industrial modernization and consumer capitalism became more and more apparent in densely populated European countries, a quest for security from the uncertainties of life under technoscientific modernization is argued to have caused a shift in the goals and political outlook of populations. Since the turn of the twentieth century, recognition of unintended side-effects has crept into social scientific discussions (Weber, 1946). These include environmental disasters or widespread medical problems. But Beck observes that unintended side-effects spawn their own *unexpected* side-effects. Risk society politics, for example, is thus all about security and the avoidance of risk. In reflexive modernization, people revisit the tenets of twentieth-century modernity and their legitimacy is questioned (Beck *et al.*, 1994: 176). Adams exaggerates for effect but conveys the general idea: individuals are 'no longer concerned with attaining something "good", but rather preventing the worst' (Adams, 1995: 182).

Recent examples of unanticipated side-effects include specific events such as the radioactive cloud spread north and west over Europe and the UK from Chernobyl, Ukraine; explosion of a chemical plant in Bhopal, India, and more general reports of global warming caused by pollutants such as ozone and sulphur dioxide, or widespread medical problems (such as increasing asthma and allergy rates which now affect children in OECD countries).[1] A 'revolution of side-effects' hits politics, communications and knowledge:

> As the risk society develops, so does the antagonism between those *afflicted* by risks and those who *profit* from them. The social and economic importance of *knowledge* grows similarly, and with it the power over the media to structure knowledge (power and research) and disseminate it (mass media). The risk society in this sense is also the *science, media and information* society.
>
> (Beck, 1992: 46)

KNOWLEDGE SOCIETIES, THE MEDIA AND RISK

Paradoxically, risk society politics arise out of more knowledgeable populations and the growth and circulation of information. The democratization of knowledge is destabilizing, as people become more and more aware of the risks taken, the 'collateral damage' of heroic national development projects and conquests, the partiality and incompleteness of scientific information, and the contingency of institutions which may claim universal authority and legitimacy but are shown to be disorganized and ill-informed or are unable to respond effectively in times of crisis (for example, in the case of government attempts to reassure the population of the safety of the beef supply chain after revelations that bovine spongeform encephalitis (BSE), had skipped from cattle to humans in the UK). Experts disagreed on the nature of the problem, institutions undermined each others' response, and politicians engaged in PR stunts which appeared foolhardy, such as one minister feeding his daughter a beef burger for the television cameras. Debates concern not only the actuality of concrete dangers, but the ways in which dangers are traced back through complex ecosystems to causes, related to broader senses of security and insecurity, and the ways in which possible risks are estimated, represented and are subject to abstraction.

The media plays a key role in this dual professional and lay knowledge system of scientific narrative, reported 'facts', interpretive journalistic commentary, urban myths and everyday life. They help to represent risks which are otherwise invisible and condense a broader sense of insecurity via iconic, signature images. On the other side of the coin, however, media do more than inform. Beck pessimistically suggests that a 'media-dependent, manipulable' public (Beck, 1997: 123) is controlled through spin-doctors employed by key industries such as the energy sector and the media's tendency for short-term 'risk fashions' in which they shift focus from issue to issue, jumbling the trivial and important, rather than offering a sustained analysis of priority items (Tulloch and Lupton, 2001: 11).

This 'jungle of interpretations and jurisdictions' is a battleground of the socially reflexive definition of risk and hazards (Beck, 1992: 112). Beck gloomily sees citizens as blind and ill-equipped to make reflexive judgements without the aid of a vanguard of media and expert activists. Other researchers, however, have found more nuance (Tulloch and Lupton, 2001: 12). The media are important interpreters of abstract notions and probabilistic reckonings of risk. Pictures and lurid reports condense and concretize what is otherwise ungraspable. However, people do have the ability to work with layered identities and interpretations within bounded contexts of everyday life in which uncertain, contending expertises must be resolved 'to' contingencies of action within praxeological frameworks which are both responsive and purposive (cf. Beck, 1992: 132). 'At some times expert systems are valued in the face of health risks such as HIV/AIDS; at others they are challenged or abandoned for more experiential, embodied and "grounded" knowledges; at yet others the offerings of both are valued in combination' (Tulloch and Lupton, 2001: 14).

One potential effect is that the legitimacy of the web of social 'settlements' is eroded. This is not so much a 'social contract' as a set of ongoing trade-offs concerning the advantages and disadvantages of, for example, urban life in advanced capitalist economies. They undergird the anonymous, aloof, commodified and 'limited liability' forms of interaction of these societies. As Tönnies noted, modern society must be sustained without the deep ties and obligations to clan and land (articulated as contracts rather than bonds of blood and soil). But the securities offered via the institutions of the family and wage labour have been eroded. In many OECD economies, full employment as a utopian goal (and means) of welfare has been abandoned. This is accompanied by increasing restrictions on access to welfare 'safety nets' and public health services. Unanticipated outcomes from industry and warfare rebound on elites and classes who believed (and often continue to believe) themselves immune or 'out of range' of environmental, political and cultural fall-out. The environmental risks of modern industry and technology (such as pollution and toxins, chemical sensitivity, radiation – Beck, 1992: 22) have shown themselves to be 'democratically' inclusive in their impact. They are invisible and unexpected in their incidence and potentially catastrophic, causing untimely deaths.

The intangibility of these types of risk confounds attempts by elites to 'buy their way out' of danger – these risks come as surprises or as a series of shocks affecting an entire population: environmental disasters such as oil spills, technological breakdown such as the collapse of electrical supply, food safety scandals, violent storms attributed to global warming and so on.

Beck argues that chains of such events produce coalitions across the class cleavages which characterized the politics of first modernity:

> the risk society can only be grasped . . . if one starts from the premise that it is always also a knowledge, media and information society at the same time – or, often enough as well, a society of non-knowledge and disinformation.
>
> (Beck, 2000: xiv)

Mass media are crucial connectors of locales which are not only physical places but information contexts interconnected in an uneven geography. This unequal knowledge system is the essence of globalization. Other circuits, including the Internet and telephone, allow experiential and firsthand accounts to move between places, to comment on, and alert others to, mass media such as televisual images or even published academic research. These circuits co-exist in a dialectical relationship which is both spatially extensive and temporally intensive – punctured by events, special occasions of intense information flows, and hiatuses and periods of silence which may amount to forgetting and disconnection (Ronell, 1989).

INDIVIDUALS AND THE EXPERIENCE OF RISK

While risks are debated at the level of expert knowledge and public accountability, they are dealt with by most people at the level of the intimate and everyday (Tulloch and Lupton, 2001: 14) within the context of family life and friendships. As modernization and globalization have proceeded, the power of states to manage risks has been lessened in the name of trade liberalization (Beck *et al.*, 1994: 131; Lash and Urry, 1994: 37ff.). Responsibility for avoiding risk comes to rest with individuals who are forced to make choices, becoming 'consumers' of risk. This spills over into the consumption of actual dangers, as in the case of death-defying extreme sports. These are all characterized by taking the body beyond its normal operating parameters as understood by modern culture (including science). Examples might include attitudinal extremes, such as defying claustrophobia and cultural attitudes towards the subterranean, or physical extremes such as ascending Everest without oxygen or experiencing zero gravity and virtually hitting the ground during a bungee jump. Risk society, then, entails not only an avoidance of danger but an active and reflexive engagement with fear.

At the same time as risk is publicly discussed in the expert terms of science and public relations terms of media-savvy corporations and campaigners, it is experienced in immediate, personal ways. Victims

> themselves become small, private . . . experts in risks of modernization. . . . What scientists call . . . 'unproven connections' are [for parents] their 'coughing children' who turn blue . . . and gasp for air with a rattle in their throat. . . . The immediacy of personally and socially experienced misery contrasts today with the intangibility of threats.
>
> (Beck, 1992: 61–62)

People are forced to make decisions which will affect their life chances and the lives of those close to them under conditions of ambiguity. Risks are invisible; many dangerous impacts come as a shock because people believed themselves to be protected. They must choose between lifestyles, self-reflexively examining the alternatives – although not just in terms of risk, as the proponents of the risk society thesis argue – but also in terms of status (why else the popularity of large vehicles such as SUVs?). How people deal with risk depends on past experience and on how it is perceived, culturally and technically. Adams (1995: 9) argues that ordinary people develop systematic ways of dealing with hazards and insecurities of everyday life.

> Employment risks, financial risks, family and relationship risks, health risks, environmental risks are all conceptualised and dealt

> with in . . . culturally distinctive ways. . . . The cultural aspects of risk perception have been treated as error in most 'scientific' approaches to calculating risk and understanding perceptions of insecurity and hazards including social science approaches such as rational choice theory.
>
> (Adams, 1995: 180ff., cited in Banks *et al.*, 2000: online)

But most have neither a methodology for making decisions rigorously, nor guidelines on how to adjudicate between conflicting information sources (the plethora of websites, manufacturers' instructions and government warnings). Urban myths may be accepted as equally plausible as scientific diagnoses and treatments. Risk calculations are difficult because of the lack of sufficient data on which to make causal interpretations of toxins and impacts and the extent to which most people are surprised by events. Problems with representing risk and complex systems in scientific terms are compounded by the everyday invisibility of environmental hazards. Even in the case of economic risks, social science delivers equivocal answers. As a result, people rely on information gathered from often untested sources and plausible inferences about cause and effect.

The constant challenge of everyday life in a risk society is to relate spaces and systems of knowledge to spaces of action which are circumscribed by the reach of personal biography. Knowledge-intensive approaches (learning organizations, knowledge management) are in part responses to the overtaking of modernity by risk. However, they generate further impacts and entail a new set of risks as they change the responsiveness of society – its ability to act.

It is difficult to make policy on the basis of evidence because there is no clear cause-and-effect chain linking policy to societal outcomes in a sociotechnical manner (Podgorecki, 1992). Authorities that promulgate plausible inferences risk their credibility if proven incorrect or if they appear implausible to others. Even the OECD admits that there are few mechanisms by which public agencies learn from the impacts of their policies because of the lack of evaluation, responsibility (staff and politicians turn over after a few months or years, leaving others to deal with problems arising from policies and programmes) and interest (OECD, 2001: 5).

The analysis of risk societies and their politics requires an emphasis on the interconnection and flows rather than breaks between different places and regions in a world, understood as a global space of distance and difference. Just at the moment when we have perhaps understood the standing apart of difference, of *différance*, most acutely, the problematic of flow and interconnection of what was previously thought to be distant, disconnected and unrelated appears as a central feature of events and problems.

The nature of knowledge changes in a risk society. Knowing shifts from being a question of analysis to a question of relations. The compositional (balance), spatial (relations in space) and deductive (relations in time) qualities of 'aesthetic reason' are more strongly figured alongside the 'categorical reason' preferred by modernists. The methods by which effective knowledge is obtained shifts. This does not eliminate difference, but demands comparative, chronotopic and simulacral methodologies which define relations in time–space, rather than merely analysis of problems into components.

A TETROLOGY OF RISK

Where risk was once merely the likelihood of danger (potential), a possible but not actualized threat, it has a complex relationship to the pursuit of security as a general objective (abstraction) and the sense or feeling of security (virtual). The category of the concrete includes both 'real and present danger' and being physically, 'actually', safe from danger. Trust that one is safe and secure is virtual. Recognition of such real but intangible attributes characterizes many of the shifts in political economic thought in the past decade. However, they are often neglected because they are misunderstood as abstractions which are not easily amenable to incorporation in economic theorems. We might summarize this, again in the form of a table (Table 8.1).

Table 8.1 Tetrology of risk and security

	Real (existing)	*Possible (non-existing)*
Ideal	Feeling of security Trust Virtual risks	'Security' in general, as a business Representations of risk as threats 'Urban myths'
Actual	Being physically safe Danger	Being 'at risk' (probability of danger) Risk

In the risk society thesis, the understanding of risk is broadened from the probable to include other ontological modes of risk. In reflexive modernization, or risk society, the conception of risk changes. Risk is not just a question of calculable probabilities but a problem of knowledge and of the relationship between the possible and the real, the virtual and the concrete.

People deftly, but in unexamined ways, package these ontological modes in linguistic signs which denote complex and often unstable blends within representations and discourses on risk. The weighting of these modes within discourse indicates the changing sociocultural understanding of particular 'risks'. The historical relations between these categories, the social forms they take, are central to cultural formations. They are integrated together in epistemic 'worldviews' and embedded in concrete institutional forms. In this sense, reflexive modernization entails not only a risk society but a new *risk culture*.

Actuality: safety and danger

Risk is understood in everyday life as danger and its absence, safety. Safety is a tangible, embodied and concrete state in many cultures. One is 'safe' from intruders (physical danger) but is never entirely safe from insecurity, paranoia or virtual risks such as magical rites (which one can typically only protect against via other magical remedies). Rather than a calculation of risk (see below), a person asks 'is it dangerous' (are there tangible dangers, perhaps falling objects on a construction site)? And, can I avoid the dangers (possibly by staying alert)?

Risk as probability

By contrast, risk as probability is founded on the regular reoccurrence of events. George Boole, writing in the 1950s, remarked on the centrality of calculation rather than emotion:

> The rules which we employ in life-assurance, and in the other statistical applications of the theory of probabilities, are altogether independent of the *mental* [i.e. virtual and abstract] phenomena of expectation. They are founded upon the assumption that the future will bear a resemblance to the past; that under the same circumstances the same [concrete] event will tend to recur with a definite numerical frequency, not upon any attempt to submit to calculation the strength of human hopes and fears.
>
> (Boole, 1958: 244–245, quoted in Porter, 1986: 81)

From the 1800s, social and cultural determinants of risk and responses to it have tended to be treated only in 'objective' and scientific/rational approaches to impact on assessment and risk management. For example, probability calculations have been used to set rates for life insurance and annuities since the first mortality tables in 1693. The application of statistical thinking and probability to the social world involved many of the most renowned mathematical and sociological thinkers of the past 200 years. Statistical and probabilistic approaches limit risk to the actual, specifically as a relationship to its two modes: the probable and the concrete. Most institutional responses have still been to dismiss broader forms of risk knowledge as a non-existing ideality; that is, as 'abstract' and entirely conceptual. Hazards are dissected and differentiated into specific, calculable risk components, allocating each to a different scientific speciality. The ontological category of 'possible actuality' (i.e. non-existing but actual; understood mathematically via probability) was developed by insurers and administrators of quarantine as a manner of quantitatively managing risk.

Risk as abstract: the efficacy of prayer

Urban myths are extended, plausible expressions of doubt and insecurity in the face of official or expert opinion and institutional reassurance. These narratives of resistance are particularly difficult to correct, not only because of their plausibility but because of their *affirmation* of the invisibility and intangibility of virtual risks and abstract threats. When we say that threat is abstract, we signal that its status is non-existing and not actual. Threats are asserted, they are a matter of perception and are feared because they may materialize as concrete danger. The intangibility of threat derives from the difficulty of calculating the odds, thereby converting it into a risk. Particularly in a situation where one does not have the benefit of historical experience or statistical summation, such as a personal encounter with a 'threatening individual', one faces a highly unstable decision line demarcating responses such as 'flight' or 'fight'.

One of the most notorious moments in the promotion of understandings of risk as entirely and only a matter of actuality was Francis Galton's 1872 article 'Statistical inquiries into the efficacy of prayer' in which he offered evidence to test whether or not prayer brought any objective advantage. Porter summarizes the argument:

> Sovereigns, whose lives were the object of regular prayerful appeal of whole nations, proved to live no longer than other members of the prosperous classes. Galton supposed that

> clergymen could be expected to plead on occasion for their own health, yet found that their life span was similar to that of physicians or attorneys. The final blow was struck by the practice of insurance companies which . . . did not distinguish between the lives of the pious and the worldly. They even offered slave vessels the same . . . rates as missionary ships.
>
> (Porter, 1986: 137)

Piety was no match for the evidence of insurance practice, but a more generous analysis might consider the very understanding of risk that prayer entails. Rather than statistically regular mishaps and disastrous events, prayer is addressed to the perpetuation of concrete states of affairs such as the fertility of crops or the arrival of a rainy season. Less often, prayer addresses the out of the ordinary: the 'bolt out of the blue', a miraculous salvation. Whatever one thinks of the evidence, where prayers are answered by miracles, the description of events is a matter of non-existing, ideal possibilities which leap into the concrete, becoming miraculously real (discussed in earlier chapters). The Judaeo–Christian and Islamic heritage draws on common notions of the miracle. Other cultures manage the relations and transformations between the abstract and concrete differently. At face value, one is inclined to agree with Galton:

> If prayerful habits had influence on temporal success, it is very probable, as we must again repeat, that insurance offices, of at least some descriptions, would long ago have discovered and made allowance for it. It would be most unwise, from a business point of view, to allow the devout, supposing their greater longevity even probably to obtain annuities at the same low rates as the profane.
>
> (Galton, 1872: 134)

However, on the other hand one routinely observes workers in the insurance industry in churches, synagogues and mosques. Furthermore, it is commonplace to resort to prayer when all else fails or when one is beyond the temporal ambit of insurance policies.

Virtuality: security and trust

The acknowledgement of the importance of trust and security as virtual aspects of risk tends to divide the social and natural sciences (Adams, 1995). Trust, typically relegated to psychology, has become an important research issue in organizational studies and in information and communication research, which has sought formal theories of this 'soft' notion (Castelfranchi and Tan, 2001: xviii).

Trust is more than secure communications, the verification of messages with senders or contractual formalities (*contra* Williamson, 1993). Lewicki and Bunker identify three types of trust: knowledge, identity and deterrence-based (Lewicki and Bunker, 1996; Lewicki *et al.*, 1998), but the list can be expanded. Trust includes:

- reliability as much as accuracy and is based on successful communication exchanges over time (knowledge-based);
- the assurances of trusted third parties (as when recommendations are made);
- the presentation of credentials or signs of legitimate membership (as when someone gives the correct password);
- forms of trust based on shared goals or identifications ('identity-based' – region, ethnicity or family, as when children trust their parents);
- the authority of legitimated institutions ('institution-based' (Castelfranchi and Tan, 2001: xxi) as when we accept the authority of a doctor based on credentials issued by a medical licensing agency);
- trust based on a social construct such as a contract which carries penalties if it is violated (deterrence-based).

These forms are not mutually exclusive – formal bonds complement what has been called informal confidence (Castelfranchi and Tan, 2001: xx; Shapiro, 1987). Motivation and disposition count towards this 'goodwill'. This is often established as part of an 'environment of trust' via extra-curricular and redundant interactions concerning matters which are not strategically important. Hence the importance of leisure activities in establishing, not a knowledge of how well a potential business partner plays golf, or exactly what movies a potential romantic partner has seen or wishes to see, but rather the cognitive and behavioural framework in which they engage in these activities. This broader framework is not reducible to an abstract concept such as 'taste' or scores, but is virtual inasmuch as it is an essence, which one expects, as a plausible inference, to inform or recur in concrete actions in the future. In a sense, this virtual knowledge provides a simulated environment in which potential actions and outcomes may be projected and evaluated.

While trust may be actualized as 'trustworthiness', its importance is that it allows actors to accept environments which are less safe in concrete terms. In other words, trust as a virtual modality of security may compensate for concrete risks. Actual outcomes in which trust is betrayed may be minimized in favour of a narcissistic preference for our own virtual and cognitive frames (Goffman, 1974). As has been argued above for risk, trust is also always more than a matter of concrete measures, formal solutions or explicit contractual bonds (Ghoshal and Moran, 1996).

Amidst the tendency towards individualization and the shift of responsibility for health and safety from the state to individuals, new forms of communities or associations arise to sustain and manage relations between modes of risk. These include clubs for leisure activities, business associations (O'Connor and Wynne, 1996) and forms of interpersonal disclosure such as the 'breakfast meeting', common in the 1980s. Anthony Giddens argues:

> This society is not only a 'risk society'. It is one where the mechanisms of trust shift in interesting and important ways.

> What can be called active trust becomes increasingly significant to the degree to which post-traditional social relations emerge.
>
> (Beck *et al.*, 1994: 186)

In some cases, risks themselves are virtual. Rather than a specific virus or other cause, symptoms are grouped together into a conventionally accepted syndrome. These tend to rise and fall historically as they are relabelled, subsumed under other diagnoses, often under the influence of available treatments and pharmaceuticals. Take, for example, the well-known – and deadly – case of 'asthma'. Asthma is a catch-all term which emerged into widespread usage in the late 1970s and early 1980s. It is defined not by particular pathogens, but by a medical consensus on when to label and how to treat a chronic and paroxysmal problem of reduced airflow to the lungs:

> Asthma is characterized by paroxysmal or persistent symptoms such as dyspnea, chest tightness, wheezing, sputum production and cough associated with variable airflow limitation and a variable degree of airway hyper-responsiveness (AHR) to endogenous or exogenous stimuli. Inflammation and its resultant effects on airway structure are considered the main mechanisms leading to the development and maintenance of asthma. . . . The definition remains valid; both airway inflammation and structural changes are still considered important in the development of clinical manifestations of asthma. There is little doubt that most cases of asthma occur as a result of environmental effects on the airways that trigger a series of modifications of the immune system in genetically predisposed individuals. In other cases, asthma may develop after toxic exposures (e.g. high level irritant-induced occupational asthma).
>
> (Boulet, 2001: online)

The essence of this consensus is a treatment regime of doses and types of drug based on past experience of patients in crisis with 'asthma attacks' and defined by the available medications. The objective is the 'management' of 'asthma', but there is no precise measurement of success in management.

However, virtual risks are predominantly relegated to communications media and 'soft' technoscience such as psychotherapy; that is, in Euro-American societies, risks are taken seriously only if they can be depoliticized. This is done by coding them in a *sub-political* discourse of expertise and technical problem-solving that corresponds with either the 'hard' technosciences or the 'iron fist' of governance such as law and the military. Risk-as-virtual, i.e. as a sense of security, as traumatic memories of negative events in the past, as anxiety and anticipation, is an ideal–real with actual impacts (stress, psychosomatic illness). This is more difficult to subject to a calculative rationality, although a person's response may draw upon the cultural heritage of the scientific method in modern thought.

RISK CULTURE OF EVERYDAY LIFE

In everyday contexts – the spaces of individual action – respondents' own understandings of risk must be produced in relation to local factors (whether a local industry or acidification of the environment), and in relation to their own status and pleasure-seeking personal behaviours (for example, sexual choices in the context of HIV/AIDS, or the decision to purchase a highly polluting model of automobile). The expansion of the 'space of worry' and of unpredictability to overlay everyday space is manifest in changed risk attitudes of the household, small town and community.

'Risk culture' has been presented above as the attitudinal and value-laden practices of reflexive modernization. This form of 'soft' infrastructure is a virtual yet important aspect of the relation to risk at two levels. At a general level for many, risk culture may be hypothesized to be a schizophrenic mixture of anxiety and insecurity concerning invisible collective risks and thrill-seeking behaviour with respect to personal danger. On a more specific scale, risk culture denotes the engagement of households and individuals with risk (for example, a decision to drink only bottled water) and the pursuit of abstract security (for example, via donations to a conflict resolution charity). As a mediating framework for decision and action, risk culture is an artefact and synthesis of the tetrology of risk, above.

This is a virtual overlay on top of the actuality of everyday life (Van Loon, 1997). The everyday is generally defined as a set of pre-reflexive practices; the domain of commonsensical, habitual and embodied thought which is practised rather than actively reflected upon. Risk culture introduces a new level of reflexivity. Proponents of the risk society thesis take the flurry of media reports in the 1980s and 1990s as indicators of the active engagement of broad sections of the populations of European and other countries with risk. We have to revise our knowledge even in – and especially in – the taken-for-granted world of everyday life. Everyday life also emerges as a newly prominent area of social science investigation (Gardiner, 1999) as it becomes more complex, and richer in the flows of information. In short, the everyday loses its innocent status.[2]

SUMMARY

Risk has become such a talked-about aspect of life in European and North American societies and in their media that a number of theorists, most prominently Ulrich Beck, have proposed that we live in a 'risk society'. This is a second, reflexive, modernity subject to the unintended impacts of expansion and the by-products of its industry. A political revolution caused by these side-effects includes debates which rage between various information sources and interpretations, including popular knowledges. Lay notions of plausible causes of danger are debated alongside expert opinion and institutionalized standards.

Risk is virtual and ideal as much as actual. Whereas risk was once dealt with in terms of *concrete* dangers and calculation of *probable* risks, the risk society thesis suggests that *abstract* notions of threat or even urban myths are also significant aspects of an overall sense of threat and crisis. *Virtual* elements such as trust and security have become key components of public assessments and decisions regarding risks. Rather than a science of insurance and risk management focused on calculable risks, a *risk culture* has evolved as a synthesis of the various modalities of risk. Insecurity, urban myths, narratives of doubt and threat, entwine with measurable events and dangers, and accounting procedures for risk. Risk culture constitutes a virtual overlay on what has long been understood as a concrete and unreflected-upon field of everyday habit and banal routine. From this we might conclude that the risk society thesis implies a shift in the constitution and conception of everyday life.

9

THE FUTURE OF THE VIRTUAL

> The speed and force of contemporary virtualization are so great that they exile beings and the attendant knowledge, alienating them from their identity, skills, and homeland. . . . Do we resist virtualization, converge on the threatened territories and identities? This would be a fatal error. . . . It is important that we try to accompany and give meaning to virtualization.
>
> (Lévy, 1998: 186)

A recognition of the virtual has been poorly developed in modern capitalist societies. The widespread use of computers and the rapid adoption of the Internet by people in many countries brought the virtual to a cultural prominence that it had not enjoyed since before the turn of the century.

- In the first chapter we argued that, for example, large-scale panoramas provided a visual form of the virtual, an early attempt at creating a virtual reality albeit a non-interactive one. The key to the understanding of the virtual developed in this book is its contrast with the concrete.

- In Chapter 2 this is made clear through an examination of the philosophical debates around the nature of the virtual as an ontological mode. The virtual is real, but not actual; ideal but not abstract.
- Moving to contemporary, digital forms of the virtual in Chapter 3, the development of simulation and virtual reality technologies, the take-up of video games and other forms of digital virtuality are argued to heighten skills and conceptual understanding of the virtual.
- The relationship of the virtual and its digital forms to globalization is clarified by considering the exclusion of most of the African continent from the networks and infrastructure necessary to support complex digital communications. Critics worry that the inequalities of colonial politics and late twentieth-century economic domination will translate into a new era of information inequality.
- Meanwhile, in the everyday lives of Europeans and Americans, information systems that rely on the virtual, and virtual meeting fora have proliferated. The balance between family time and commercial interruptions, work and home life has been decisively shifted away from the family (Chapter 5).
- Workplaces and workforces are now virtualized, in the form of flexible, work-from-home and on-call arrangements (Chapter 6).
- The spectacular business failures discussed in Chapter 7 demonstrate that there has been insufficient attention to managing the relation between the virtual and the concrete.
- Looking back, if, in Chapter 1, we saw Archbishop Cranmer lose his life in 1555 over his assertion of the virtuality of the Christian Eucharist, in later chapters we observed the role of the virtual in insecurity (Chapter 8) and the death of corporations such as Enron for their risk-taking.

The virtual implies a willingness to believe in the reality of dreams, and marks the concern with history and the past as well as creative change. Dyens (2002: 33) agrees that the virtual is 'a perception that is alive' which forces us to re-examine our ontological assumptions, 'defined by a biological understanding of the world' that only organic matter is alive or participates in social interactions (see also Chapter 2). It furthermore challenges our epistemological assumptions and 'truth practices' in which we tend to treat the concrete as the only and final site of 'Truth'. In the case of digital virtuality, rather than being concerned about the immateriality of the virtual it is the exchange between organic matter and the virtual or cultural that is of paramount importance. Dyens holds out the possibility of conceiving of a 'cultural animal', a 'non-organic being' that fuses the concrete and virtual. Such a being might consist mostly of information – such as a virus (a parasitical sequence of DNA or RNA, which uses a host organism to reproduce and disseminate itself, which has until recently been treated as non-living but organic, informational but biological – in short virtual, (Dyens, 2002: 33ff.).

After the self-preoccupation of the 1960s and 1970s, late twentieth-century Europe and North America faced the future under a curtain of uncertainty. An overriding concern was with the past: intellectuals and spokespersons for the oppressed advertised historical injustices, all of Europe became preoccupied in one way or another with the Balkans' fealty to ancient ethnic and regional ties, attention was given over to the significance of the wars of the first half of the twentieth century and commemorating their victims and combatants. Even the end of the twentieth century was compared to past *fin de millennia* and ancient apocalyptic prophecies.

Revivals and fundamentalisms underline the Christian and Muslim message of a divinity and a heaven which was not abstract or in some way outside of the ontological schemes that humanity might comprehend but, real, and if not physically present, then virtually present. With these movements, secular Utopias went to Heaven. The virtual entered politics. The abstract goals and principles of earlier proclamations of human rights, utopian faith in technology, evaporated into presentism, the endless circulation of novelty (Lyotard, 1993; Maffesoli, 1996) and a search for values which were found in the tradition and pragmatism of everyday life.

Fundamentalisms stress the virtual because of their insistence on the reality of the ineffable – God exists. Holy texts are to be read as literally true: they are accurate accounts of the past. By contrast, many liberal Christians during the twentieth century in particular slipped into an understanding of the Bible, in particular, as a literary work, an approximate tale, to be taken 'with a grain of salt'. Thus many doubted the accuracy of accounts and allowed only that some parts may be historically true (i.e. virtual). Religious 'tall tales' of miracles were treated as never-existing abstractions, fancies of the mind. This gave rise to archaeological expeditions and linguistic reconstructions in an attempt to 'discover' the historical Jesus or events and locations in the Scriptures. Although the steady popularity of fiction and novels suggests that we have hardly abandoned the abstract, critics warn that the spread of simulations and visualization technologies which rely on a normalization and routinization of the virtual as part of our conceptual toolkit at work and at home may mean that our willingness to put time and effort into imagining anything:

> By shifting to a virtual world . . . we move into a world where everything that exists only as idea, dream, fantasy, utopia will be eradicated, because it will immediately be realized, operationalized. Nothing will survive as an idea or a concept. You will not even have time enough to imagine. Events, real events, will not even have time to take place. Everything will be preceded by its virtual realization. We are dealing with an attempt to construct an entirely positive world, a perfect world, expurgated of every illusion, of every sort of evil . . . exempt from death itself. This pure, absolute reality, this unconditional realization of the world – this is what I call the Perfect Crime.
>
> (Baudrillard, 2000: 66–67)

However, as Bogard points out, the virtual entails a world full of different forms of simulation, all of which attempt to control and programme the future (see Chapter 3; Bogard, 1996). Power moves into an anticipatory, future-oriented mode. Entertainers, politicians, corporations, bureaucracies attempt to anticipate demand and desires. Publics demand that law enforcement anticipate and prevent crime (for example through surveillance cameras), a very different activity than the original purpose of the police, which was to investigate, prosecute crime and to supervise the punishment of criminals. By implication, resistance and opposition must also move out of the 'concrete' mode with its stress on reaction and demonstration in the present towards a future-oriented mode geared around issues that may not yet have spawned disasters or events. However, the experience of global environmental change and movements for global economic justice illustrate how difficult it is to use probability to persuade leaders and publics (for example, for over a decade the actuality of global warming was doubted and causes are still disputed). Specific, localized events and actions become the concrete campaigns which actualize the virtuality of far broader claims such as global warming. The hope is to mobilize popular sentiment in favour of the larger cause, not just a specific issue (for example, the battle over logging around Claquoit Sound on the west coast of Canada).

TERRORISM AS VIRTUAL WAR

On 11 September 2001 suicide teams simultaneously hijacked four flights. After slaughtering the crews, one crashed, a plane was flown into the Pentagon, and the remaining two were flown into each of the towers of New York's World Trade Center. The attacks were a *jihad* in the name of Islam and the *Al-Qaeda* terrorist network. Their spectacular collapse entombed thousands of global financial and commercial workers. 'Live-on-TV' coverage brought a dark side of globalization home to viewers. Traumatized by the actual carnage of the collapse of the World Trade Center towers, survivors exclaimed that it was 'like something on TV', a cinematic nightmare, a virtual reality simulation come alive. In the initial confusion some hoped that they were witnessing a giant televised hoax, a fantasy which would not intrude on the concrete world of people and their lives. But for many there was a strange quality of the virtual becoming real. The images resembled a Hollywood disaster movie. Many had played 'Flight Simulator' software in which New York is represented only by the twin towers of the World Trade Center. Since 1990, children routinely crashed their simulated aircraft into the towers to see what would happen next in the game (depending on the selected point of view, a cracking glass windscreen is depicted).

In the hearts of those who sacrifice themselves for their ideals, the virtual reigns over the material. Obstacles, costs in terms of lives – nothing matters but 'The Cause', Heavenly redemption, 'righteous' justification. Terrorism is in effect a violent form of communication which uses the lives of its victims as a message of intimidation to the rest of the population. It brings the paranoia of film noir to the innocent spaces and times of everyday life. But unlike past forms of terrorism, which were generally accompanied by explicit demands, the aim of these terrorists seems simply to divide populations, exacerbate religious conflict between Muslims and non-Muslims, and to provoke war between states.

The subsequent US bombing of Afghan cities mirrored the terrorist action, dispensing violence from the skies against cities and civilians. It appeared to many commentators that the USA and its allies have become trapped in a script written by the terrorists. Actual life seemed to have become a strange echo of the virtual – of a pre-ordained response, of the the virtuality of metaphysical beliefs. Such attacks punctuated the rhythm of everyday life and the schedules of global capitalism to remind secular 'Westerners' of the virtual. When 'virtualities' are enacted, made actual, they are usually buffered with ritual.

When we are caught unprepared for this, we momentarily lose our bearings as all that is solid melts before memory and images. The war in Afghanistan and the anti-terrorist campaign globally is a war over virtualities, over 'real ideals'. But all wars are fought in the actual, with material consequences.

In the 1990s, Jean Baudrillard notoriously argued that the Gulf War never really happened because it took place only on the screens of television viewers and of pilots in fighter jets. The virtuality of the Gulf War which Baudrillard sought to highlight was of course only one side of an equation in which bombs actually fell and caused deaths, material destruction and devastated a population. Many of the deaths were those of children, casualties as much of cynical warring states as of bombs. Iraq calculates that, in suffering such losses, its people will hate the USA and everything it stands for. The US Pentagon calculates that Iraqi families will blame the regime in power for the loss of their children. If the Gulf War was virtual, it was not so simply for technical reasons but because of this overlay of terroristic policies. Virtual war is terrorism.

THE FUTURE OF THE VIRTUAL

We need to know more about the virtual as it is the necessary philosophical category of all innovation. It is not to be confused with the abstractly possible (the abstract). Social change, emergence, or the unfolding of processes does not merely realize an identity already present in abstract concepts, as in the realization of a plan. Actualization is closer to performance where each instantiation is unique rather than being a copy. The virtual provides a 'handle' for philosophers to designate the coming into being of difference, of processes of change, and the many forms that the concretely actual, for example firms or everyday interactions, can take.

What is the future of the virtual as an aspect of social life? The virtual is essential to understanding the increasing weight with which we feel, and must count in, absences. 'Out of sight, out of mind' is a maxim that speaks to the actual (which continues to rule the parochial). As a category, the absent has therefore been conceived of as abstract, as a *non-existing ideation*, giving rise to a philosophical problem of how to speak about absence except oxymoronically as non-presence. The discussion of globalization in Chapter 4, however, led to the conclusion that the virtual is the true category by which absent, distant decisions that impact on local routines may be understood as real, although not materialized in a given, parochial context. The presence of absence is virtual.

If digital virtuality does represent continuity from the more general, historical forms of the virtual, will it be possible to speak of a 'virtual society'? Traditional virtualities may come to the fore, but a key shift may be the even stronger foregrounding of brand identities and the importance of corporate images as virtual property. Corporations, of course, are virtual persons, both subject to and in other ways beyond the reach of the law because they feel neither actual pain nor remorse.

In the virtual society, image is king. There is a noticeable investment in the rhetoric of the 'virtual society' including corporations such as Mitsubishi and Sony (see e.g. *http://www.vs.sony.co.jp*). This also appears to be the case in North America, despite the European observation that, 'there is an uneasy fit between the rhetoric of virtuality and the day-to-day problems of running an organisation' (Hughes, 1998). Consumers will increasingly recognize and be willing to pay for branded products, as not only guarantors of quality but of future resale value, such as might be the case for houses in particular 'desirable neighbourhoods' where the locale is a type of branding.

But the ineffable status of such brands raises the possibility that the virtual society will also be a society of virtual responses to concrete dilemmas: consumers who buy to improve not only their image but to cheer themselves up and to attempt to mask material ills. For organizations and governments, the virtual may equally figure in a society of disinformation, of dark media campaigns of slurs and innuendo aimed at damaging the virtual aspects of people or companies, including not only brands but charisma. And, where every action, every concrete situation and object have a prominent, virtual aspect, every action and object takes on greater symbolic importance as if part of a ritual. They are over-lit with mythic qualities and brand identity almost to the point of paranoia.

In such a situation, symbolic action takes on intense importance and may achieve the same ends as more costly and actual physical actions. Many critics share the worry of Donna Haraway:

> For William Gibson (1986), cyberspace is 'consensual hallucination experienced daily by billions. . . . Unthinkable complexity.' Cyberspace seems to be the consensual hallucination of too much complexity, too much articulation. It is the virtual reality of paranoia, a well-populated region in the last quarter of the Second Christian Millennium. Paranoia is the belief in the unrelieved density of connection, requiring, if one is to survive, withdrawal and defense unto death. The defended self re-emerges at the heart of relationality. Paradoxically, paranoia is the condition of the impossibility of remaining articulate. In virtual space, the virtue of articulation – i.e., the power to produce connection – threatens to overwhelm and finally to engulf all possibility of effective action to change the world.
>
> (Haraway, 1992: 325)

We may find that no one takes responsibility in a virtual society. It is unfashionable, and elites and leaders avoid it. Deflecting blame becomes a skill, a 'conceptual theatre of self-interest' (Kingston, 2002: SP1). Although excuses and shifting blame have been common in political life for eons, the case of Enron (Chapter 7) illustrates the tendency to both search for scapegoats and to refuse responsibility. Recently a reporter on the *New York Time Magazine* was found to have invented a supposedly real character who formed the centre of a sensational story about slavery on the Ivory Coast. The editors denied responsibility for publishing the story without checking the facts. The journalist 'tried to off-load his breach of journalistic ethics by suggesting that he was trying to forge a new form of journalism of the kind, presumably, in which reported fact and stuff you make up are given equal weight' (Kingston, 2002: SP8).

We need to know more about how 'the virtual' becomes a template for reacting to material events in everyday life. We need to know more about how the biases and conceptual categories of the past are translated into the new, virtual, matrices of the present and future. We need to learn more from past manifestations and forms of the virtual so that we understand better the implications of our ongoing investment in creating a global, digital virtuality.

Notes

1 THE RETURN OF THE VIRTUAL

1 Although in recent media-stunts people attempted to spend a year during which they purchased all the necessities of life via online shopping sources, but even here, they did not remain logged on as participants in an online, virtual environment for the whole period. The Web remained a temporary communication and logistical space. They merely directed their consumer spending to retail sites on the World Wide Web.

2 THE VIRTUAL AND THE REAL

1 The virtual is not a form of The Real in Lacan's sense. Whereas Lacan counterposes an unrepresentable actuality as an absent fullness which figures in language as excess and lack, but which can never be adequately represented. It therefore constantly troubles and undermines the authority of, for example, representations of self-identity (see Zizek, 1989; Widder, 2000: 118).

2 Debates about the world are intractable when posed in purely philosophical terms. The social and natural sciences developed out of natural philosophy as a result of the need to probe the material world about us rather than attempting to provide a purely philosophical answer. This approach considers people, machines, media and nature to be part of an integrated environment. It will be continued here. Rather than defining the virtual by contrasting it with the real, this book takes a different tack, by surveying the ways in which the term 'virtual'

has come to prominence in contemporary usage, and examining the usage of the term itself to see what it means to different people.

3 Some worry that the increasing prominence of the virtual represents not only a shift from the actual to the ideal, as discussed in this book, but a corresponding shift towards an imbalance away from abstraction in favour of virtualities. In the terms set out below, that is to say a shift from the possible towards a greater bias in favour of the real. This could entail a loss of the free-wheeling creativity afforded to the imagination to create purely conceptual abstractions (concepts, fantasies) (see Baudrillard, 2000: 66).

4 Hegel provides a theory of the realization of this abstract Idea into the expression of a collective Spirit in the form of the political nation-state. His theory of dialectical negations is an attempt to visualize the historical and political process by which identity is realized through a process of negation that specifies differences between identities. Development continues in attempts to surmount – by negation – the contradictions that have been created between them. The initial positive state is an abstraction because it is 'indifferent' to any other object but it therefore cannot be specified or even accounted for (Hegel, 1977 ss. 578 in Widder, 2000: 124; Rorty, 1967).

5 If one struggles to put meaningful terms into the matrix that can be created from Proust's definition, the expected correspondences found in, for example, Bergson and Deleuze do not appear. Deleuze's suggested arrangement leaves one floundering to conceive of a 'possible', that is, a non-existing virtuality. He himself argues this is an impossibility, a null-set. Sketching in the 'real and actual' as the concrete (in square brackets) produces a Platonism in which the real includes both ideal forms and concrete objects (Badiou, 2000). The abstract, understood as a transcendental that exists only in concept but not in reality, fits poorly but is needed to avoid a form of Platonism. The concrete can easily be lost altogether.

6 However, he is criticized for constructing a philosophical system in which that amounts to the same thing (Badiou, 2000 – see n. 1 above).

7 A flaw in Russell's empiricism and Capor's positivism disproved by Karl Popper.

8 Such realist positions require the category of the virtual to avoid the accusation of committing a *post hoc* ergo *procter hoc* fallacy of conventionalism – that they test only for the existence of the intangible structures they assert rather than searching for simpler explanations.

3 DIGITAL VIRTUALITIES

1 In digital domains, the network, with all its computing and telecommunications infrastructure, the conventions of digital addressing and of data processing, precedes virtual space in an even more literal manner than Baudrillard could have dreamed of when he remarked, 'the map proceeds the territory' (Baudrillard, 1990: 1). Even the current notion of the website is a gloss on what is a strictly codified manner of retrieving and displaying data. Web pages themselves are composed out of linked elements such as graphic image files, punctuated by hypertext links to other data and files. Hypertext links are indexes caught on the threshold of departure, signalling to another page or text. They are paradoxical because they appear to be an interior gateway. To indulge in an architectural metaphor, a link is less a portal to the outside and more like a hidden passage in a building – a door to the inside, that leads out somewhere else, reinforcing the sense of self-sufficient totality achieved in the Net. Ambiguity thus becomes 'mystery' in the absence of a span across clear categorical divisions (in this case, distinctions such as inside and outside, here and there, break down – see Shields, 2000).

2 Myron Krueger provides an early history of the development of technologies such as the light pen for sketching or selecting items on a video screen, the mouse and the graphical interface of the Apple computer, and the head-mounted display at military-funded research institutes and university labs (Krueger, 1991). Rob Kitchen's *Cyberspace: The World in the Wires* provides one of the first histories of the development of virtual reality and virtual environments, specifically relating them to the parallel but earlier emergence of the Internet (Kitchin, 1998: 45ff.) and offering a rare history of European and Japanese work (see Timeline). Hillis (1999) integrates sources to give a critical history of virtual reality in the United States and its predecessors from the pin-hole camera and Cinerama to the development of early head-mounted displays (HMDs).

3 Like terrifying monsters, the dinosaur figures large in the American imagination of a ferocious otherness, so foreign as to be locked in another space (a lost land or island) or time (the cretaceous, or the far future). Whereas wheeled transport is far more efficient, the fantasy of the walking avatar or robot is deeply ingrained in VR. Walking upright like cinematic depictions of the tyrannosaurus (or some *Star Wars* battle machine) is a mark of the human and hints at a horrifying otherness – a reptilian intelligence or cunning.

4 VIRTUAL AFRICA

1 The high is provided by the Nua Internet Survey for August 2001. The low figure is Jensen's estimate for early 2000 (Jensen, 2000). A key website is the International Development Research Council of Canada's Acacia Program: *http://www.idrc.ca/acacia*. IDRC has proposed a measure based on Bits of Information transferred per capita (BPC). Misleadingly high numbers in the Seychelles, Cape Verde Islands and Djibouti reflect foreign military presences and satellite ground stations. Other high bandwidth and relatively high scores of approximately 5 to 7 BPC are found in Egypt (highest), South Africa, Tunisia, Botswana, Gabon and Senegal. Most countries are under 1 BPC with the lowest figures found across the countries of Central Africa and the Sahara-Sahel such as Mali, Niger, Chad, Centrafrique and Democratic Republic of Congo (June 2002 figures from *http://www.idrc.ca/acacia*). In January 2000, only 500,000 Africans subscribed to formal dial-up accounts with African Internet Service Providers (Jensen, 2000).

2 Excluding South Africa, the World Bank 1999 figures list the per capita GDP for Sub-Saharan Africa at an average just over US$300 with life expectancy falling to less than 47 years due to AIDS and civil wars. By contrast the per capita GDP for the United States is US$33,000.

3 One South African electricity company, Eskom Enterprises is in the midst of a ten-year project to run power and optical cable transmission lines through the centre of Africa, north from Cape Town to Cairo.

4 For example, the case of Guo Qinghai convicted of inciting subversion in Cangzhou near Beijing in 2000 after posting an article calling for political reforms on websites overseas (Information Centre for Human Rights and Democracy, Hong Kong). Saudi Arabia blocks more than 400,000 sites according to *Al-Eqtisadiah*, a business daily cited by Agence France Presse in 2000.

5 JOYSTICK GENERATION: CYBERPUNKS, CAMKIDS AND FAMILY LIFE

1 Yet by contrast, corporations and banks have successfully protected the privacy of their methods for evaluating clients or making credit decisions (Viera, 2002).

2 The practical goal in secure commerce is not the elimination of risk but raising the 'cost' of cracking systems above the benefits reaped by a hacker. For example, as we move above 150-character encryption keys we enter a realm of very expensive computing power if the system is to be cracked, especially if it is to be done so in sufficient time to make

use of the information. If it takes a criminal three months to crack an encryption key, but the key itself is changed every week, the system may be considered to be reasonably secure (Banks *et al.*, 2000).

6 WORK: VIRTUAL WORKING

1 These comments derive from research on machine-shop and sheet-metalworkers in the construction industry in Canada (Shields and West, 2000), and cement block-pressing and manufacturing in the UK (Shove, 1996).
2 At the East Carolina School of Medicine in Greenville, North Carolina.
3 These machines include Intuitive Surgical Inc's 'da Vinci', and 'Zeus', a system designed by Computer-Motion Inc to perform minimally invasive coronary operations.
4 The Computer Museum History Center provides a well-illustrated time-line of the development of various aspects of computing up to 1990, online at: *http://www.computerhistory.org*.
5 All figures for the USA. The survey shows an increase of 17 per cent annually but includes occasional teleworkers and part-time workers from home (International Telework Association Council 'Telework America 2001' survey). Online: *http://www.telecommute.org*. European information available online: *http://www.emergence.nu*.

8 RISK, TRUST AND THE VIRTUAL

1 Rates for paediatric asthma vary between 7 per cent (in the best cases such as Denmark) and almost 30 per cent (in the worst cases such as Canada) across countries with the highest per capita GDPs.
2 I am indebted to Joost Van Loon for his suggestions on this point.

References

Abbate, J. (1999) *Inventing the Internet*. Cambridge, Mass.: MIT Press.
Adams, J. (1995) *Risk*. London: UCL Press.
Aftab, P. (2001) *The Parent's Guide to Protecting Your Children in Cyberspace*. New York: McGraw-Hill.
Agamben, G. (1998) *Homo Sacer: Sovereign Power and Bare Life*. Palo Alto, CA: Stanford University Press.
Albrow, M. (1990) *Max Weber's Construction of Social Theory*. London: Macmillan Education.
Amazon.com. (2002) *Highlights of Fourth Quarter and Fiscal 2001 Results*. Retrieved 14 February from the World Wide Web *http://www.iredge.com/IREdge/site/002239/builtin/Q4_Release_832.htm*.
Amis, M. (1982) *Invasion of the Space Invaders: An Addict's Guide to Battle Tactics, Big Scores and the Best Machines*. London: Methuen.
Anderson, B. (1983) *Imagined Communities*. London: Verso.
Anon. (2002) 'Paypal'. *Financial Post*, 16 February, p. 1.
Antze, P. and Lambek, M. (eds) (1996) *Tense Past*. New York: Routledge.
Appadurai, A. (1996) *Modernity at Large: Cultural Dimensions of Globalization*. Minneapolis: Minnesota University Press.
Badiou, A. (2000) *Deleuze* (trans. L. Burchill). Minneapolis: University of Minnesota Press.
Bakhtin, M. M. (1981) 'Author and hero in aesthetic activity'. In V. Liapunov and M. Holquist (trans.), *Art and Answerability*. Austin: University of Texas Press, pp. 4–256.
Balsamo, A. (1995) 'Signal to noise: on the meaning of cyberpunk

subculture'. In *Communication in the Age of Virtual Reality*. Hillsdale, NJ: Lawrence Erlbaum, pp. 347–368.
Banks, M., Lovatta, A., O'Connor, J. and Raffob, C. (2000) 'Risk and trust in the cultural industries'. *Geoforum*, 31 (4), 453–464.
Barbrook, R. (1996) *HyperMedia Freedom*. Retrieved July 2001 from the World Wide Web: *http://www.ctheory.com/global/ga101.html*.
Barge, J. (1994) 'Keyboard cases go to trial: new evidence shows defendants admitted injuries'. *ABA Journal*, 80, 30.
Barker, K. and Christensen, K. (eds) (1998) *Contingent Work: American Employment Relations in Transition*. Ithaca, NY: IRL Press.
Barley, R. and Orr, J. E. (eds) (1997) *Between Craft and Science: Technical Work in US Settings*. Ithaca, NY: IRL Press.
Barlow, J. P. (1996) *A Declaration of the Independence of Cyberspace*. Retrieved July 2001 from the World Wide Web: *http://www.eff.org/~barlow/Declaration-Final.html*.
Bartnatt, C. (1995) *Cyberbusiness: Mindsets for a Wired Age*. Chichester: Wiley.
Baudrillard, J. (1990) *The Precession of Simulacra*. New York: Zone Books.
Baudrillard, J. (2000) *The Vital Illusion*. New York: Columbia University Press.
BBC (1998) *Salmonella Remains a Threat*. BBC News. Retrieved 26 April 2002 from the World Wide Web *http://news.bbc.co.uk/hi/english/health/newsid_253000/253554.stm*.
Beck, U. (1992) *Risk Society: Towards a New Modernity*. London: Sage.
Beck, U. (1997)
Beck, U., Giddens, A. and Lash, S. (1994) *Reflexive Modernization. Politics, Tradition and Aesthetics in the Modern Social Order*. Palo Alto, CA: Stanford University Press.
Beckert, J. (1999) *Economic Action and Embeddedness: The Problem of the Structure of Action*. Berlin: Free University of Berlin John F. Kennedy Institute.
Beineix, J-J. (1999) *Otaku: Les Enfants du virtuel*. Paris: Denoel.
Benedikt, M. (ed.) (1991) *Cyberspace, First Steps*. Cambridge, Mass.: MIT Press.
Bergson, H. (1988) *Matter and Memory* (trans. N. M. Paul and W. S. Palmer). New York: Zone Books.
Beverley, J. (1993) *Against Literature*. Minneapolis: University of Minnesota Press.
Bey, H. (1985) *Temporary Autonomous Zone*. New York: Autonomedia.
Biocca, F. (1992) 'Virtual reality technology'. *Journal of Communication*, 42 (4), 23–72.
Blackwell, T. (2002) 'Net hotline will allow wives to report "concerns"'. *National Post*, 27 February, p. 7.

Bobick, A. F. (1999) 'The kids room: a perceptually-based interactive and immersive story environment'. *Presence-Teleoperators and Virtual Environments*, 8 (4), 369–393.

Bogard, W. (1996) *The Simulation of Surveillance: Hypercontrol in Telematic Society*. New York: Cambridge University Press.

Boole, G. (1958) *An Investigation of the Laws of Thought on which are Founded the Mathematical Theories of Logic and Probabilities*. New York: Dover Press.

Boulet, L-P. *et al.* (2001) 'What is new since the last (1999) Canadian Asthma Consensus Guidelines'. *Canadian Respiratory Journal*, 8 (2).

Bray, H. (2001a) 'A $1.8b ring around Africa'. *Boston Sunday Globe*. 22 July, p. A25.

Bray, H. (2001b) 'Africa struggles to bridge digital divide'. *Boston Sunday Globe*, 22 July, pp. A1, A24–25.

Brooks, P. N. (1992) *Thomas Cranmer's Doctrine of the Eucharist* (2nd edn). London: Macmillan.

Butler, J. (1993) *Bodies that Matter*. New York: Routledge.

Callon, M. (1998) *Laws of the Markets*. Oxford and Malden, Mass.: Blackwell/*Sociological Review*.

Cappeliez, S. (2001) *'McDonaldization'? Food in a Globalized Culture*. Ottawa: Carleton University Press.

Carrier, J. (1998) 'Abstraction in Western economic practice'. In J. Carrier and D. Miller (eds), *Virtualism*. New York: Berg, pp. 25–49.

Castelfranchi, C. and Tan, Y-H. (eds) (2001) *Trust and Deception in Virtual Societies*. Norwell, Mass.: Kluwer Academic.

Caygill, H. (1999) 'Meno and the Internet: between memory and the archive'. *History of the Human Sciences*, 11 (4), 1–11.

Citizen (2001) 'Computers not to blame for carpal tunnel syndrome: study'. *Ottawa Citizen*, 12 June, p. A12.

Colebrook, C. (1999) 'A grammar becoming: strategy, subjectivism and style'. In E. Grosz (ed.), *Becomings*. Ithaca: Cornell University Press: 117–140.

Colombat, A. P. (1999) 'Deleuze and the three powers of literature and philosophy'. In I. Buchanan (ed.), *A Deleuzian Century?* Durham, NC: Duke University Press, pp. 199–217.

Conlin, S. (2001) 'Dreamcast: Sega smash pack Vol. 1 Review'. *Ottawa Citizen*, 18 June, p. B7.

Coyne, R. (1999) *Technoromanticism: Digital Narrative, Holism and the Romance of the Real*. Cambridge, Mass.: MIT Press.

Cranmer, T. (1846) *Miscellaneous Writings and Letters of Thomas Cranmer, Archbishop of Canterbury, Martyr, 1556*. Cambridge: Parker Society.

Crary, J. (1992) *Techniques of the Observer: On Vision and Modernity in the Nineteenth Century*. Cambridge, Mass.: MIT Press.

Cubitt, S. (1998) *Digital Aesthetics*. London, Thousand Oaks, New Delhi: Sage.
Cuthell, J. (2002) *Virtual Learning: The Impact of ICT on the Way Young People Work and Learn*. Aldershot, Hants: Ashgate.
Dean, J. (1998) *Aliens in America*. Ithaca, NY: Cornell University Press.
Dean, J. (2000) 'Webs of conspiracy'. In A. Herman and T. Sloop (eds), *The World Wide Web and Contemporary Cultural Theory*. New York: Routledge, pp. 61–76.
Deleuze, G. (1981) *Difference and Repetition*. New York: Columbia University Press.
Deleuze, G. (1986) *Cinema* (Vol. 1). London: Athlone.
Deleuze, G. (1988) *Bergsonism* (trans. H. Tomlinson and B. Habberjam). New York: Zone Books.
Deleuze, G. (1993) *The Fold*, trans. Tom Conley. Minneapolis: Minnesota University Press.
Deleuze, G. (1994) *Difference and Repetition* (trans. P. Patton). New York: Columbia University Press.
Dery, M. (2001) 'Bit rot'. In *Crypto Anarchy, Cyberstates, and Pirate Utopias*. Cambridge, Mass.: MIT Press.
Dorr, A. (1983) 'No shortcuts to judging reality'. In *Children's Understanding of Television: Research on Attention and Comprehension*. New York: Academic Press, pp. 199–220.
Douglass, P. (1992) 'Deleuze's Bergson: Bergson redux'. In F. Burwick and P. Douglass (eds), *The Crisis in Modernism: Bergson and the Vitalist Controversy*. Cambridge: Cambridge University Press.
Downey, G. (2001) 'Virtual webs, physical technologies and hidden workers: the spaces of labour in information Internetworks'. *Technology and Culture*, 42 (1), 209–235.
Dyens, O. (2002) *Metal and Flesh: The Evolution of Man: Technology takes over* (Evan Bibbee and Ollivier Dyens, Trans.). Cambridge, MA: MIT Press.
Dyson, E. (1998) 'The end of the official story'. *Brill's Content*, 50–51.
Earle, N. and Keen, P. (2000) *From .com to .profit: Inventing Business Models that Deliver Value and Profit*. San Francisco, Calif.: Jossey-Bass.
Economist, The (1999) 'Mbeki's words of website wisdom'. November 13th, pp. 46.
Economist, The (2000a) 'Survey: e-management'. *The Economist*, 356 (8196), pp. 68ff.
Economist, The (2000b) 'Survey: on-line finance: the virtual threat'. *The Economist*, 20 May, pp. 66ff.
Economist, The (2001a) 'E-strategy brief: Enron'. *The Economist*, 30 June to 5 July.
Economist, The (2001b) 'E-strategy brief: Merrill Lynch'. *The Economist*, 9–15 June, pp. 79–80.

Economist, The (2001c) 'E-strategy brief: Siemens'. *The Economist*, 2–8 June, pp. 77–78.

Economist, The (2001d) 'E-strategy brief: GE'. *The Economist*, 19–25 May, pp. 75–76.

Economist, The (2002) *Everyman, Farewell* (10 January). Retrieved 19 February from the World Wide Web *http://economist.com*.

Edwards, P. (1998) 'Y2K: millennial reflections on computers as infrastructure'. *History and Technology*, 15 (1), 7–29.

Elmer, G. (1998) 'Diagrams, maps and markets: the technological matrix of geographical information systems'. *Space and Culture – Theme Issue on Habitable Spaces*, 3, 49–65.

Elmer, G. (ed.) (2002) *Critical Perspectives on the Internet*. Boulder, CO: Rowman & Littlefield.

Eyerman, A. (2001) *Women in the Office, Transitions in a Global Economy*. Toronto: Sumach Press.

Fedler, F. (1989) *Medea Hoaxes*. Ames: Iowa State University Press.

Fine, B. (1995) 'From political economy to consumption'. In D. Miller (ed.), *Acknowledging Consumption*. London: Routledge, pp. 127–163.

Fitting, P. (1991) 'The lessons of Cyberpunk'. In C. Penley and A. Ross (eds), *Technoculture*. Minneapolis: University of Minnesota Press.

Foxe, J. (1877) *Acts and Monuments*. London: n.p.

Franda, M. (2002) *Launching into Cyberspace: Internet Development and Politics in Five World Regions*. London: Lynne Rienner Publishers.

Franklin, S., Lury, C. and Stacey, J. (2000) *Global Nature, Global Culture*. London: Sage.

Friedberg, A. (1993) *Window Shopping: Cinema and the Postmodern*. Berkeley: University of California Press.

Galloway, A. (2002) *Shaping Invisible Spaces: Wireless Networks and Ubiquitous Computing* (Draft). Ottawa: Carleton University.

Galton, F. (1872) 'Statistical inquiries into the efficacy of prayer'. *Fortnightly Review*, n.s., 12, 125–135.

Gardiner, M. (1999) *Critiques of Everyday Life*. London: Sage.

George, S. and Sabelli, F. (1994) *Faith and Credit*. Harmondsworth: Penguin.

Ghoshal, S. and Moran, P. (1996) 'Bad for practice: a critique of the transaction cost theory'. *Academy of Management Review*, 21 (1), 13–47.

Gibson, W. (1984) *Neuromancer*. New York: Ace.

Giddens, A. (1990) *The Consequences of Modernity*. Cambridge: Polity Press.

Goffman, E. (1974) *Frame Analysis*. Cambridge, Mass.: Harvard University Press.

Graddol, M. (1997) *The Future of English*. London: British Council.

Green, E. and Adam, A. (eds) (2001) *Virtual Gender: Technology, Consumption and Identity*. London: Routledge.

Grosz, E. (1999) 'Thinking the new'. In E. Grosz (ed.), *Becomings*. Ithaca: Cornell University Press: 15–28.
Grover, V. and Kettinger, W. J. (1997) 'The impacts of business process on organizational performance'. *Journal of Management Information Systems*, 14 (1), 9–12.
Guattari, F. (1992) *Chaosmose*. Paris: Galilee.
Gunter, B. (1998) *The Effect of Videogames on Children: The Myth Unmasked*. Sheffield: Academic Press.
Hakken, D. (1999) *Cyborgs@Cyberspace: An Ethnographer Looks to the Future*. New York: Routledge.
Hale, R. and Whitlam, P. (1997) *Towards the Virtual Organization*. London: McGraw-Hill.
Halkes, P. (1999) 'The *Mesdag Panorama*: sheltering the all-embracing view'. *Art History*, 11 (1), 83–98.
Halkes, P. (2001) *Aspiring to the Landscape*. Unpublished Ph.D. Dissertation, University of Leiden.
Halper, M. (1998) *Making Money on the Web* (15 January). Retrieved 24 April 2002 from the World Wide Web *http://www.cio.com/archive/enterprise/011598_money_content.html*.
Haraway, D. (1992) 'The promises of monsters: a regenerative politics for inappropriate/d others'. In L. Grossberg, Cary Nelson and Paula Treichler (eds), *Cultural Studies*. New York: Routledge, pp. 295–337.
Hardt, M. (1993) *Gilles Deleuze. An Apprenticeship in Philosophy*. Minneapolis: University of Minnesota Press.
Harvey, P., Green, S., Lury, C. and Brown, S. (2000) *Network, Scale, Memory and Play: Key Concepts in ICT Research*. Paper presented at the ESRC Virtual Society, University of Manchester.
Hayes, D. (1989) *Behind the Silicon Curtain: The Seductions of Work in a Lonely Era*. Boston, Mass.: South End Press.
Hayes, R. D. (1995) 'Digital palsy: RSI and restructuring capital'. In *Resisting the Virtual Life*. San Francisco, Calif.: City Lights Books, pp. 173–180.
Hayles, N. K. (1993) 'Virtual bodies and flickering signifiers'. *October*, 66, 69–91.
Hayles, N. K. (2000) 'The condition of virtuality'. In P. Lunenfield (ed.), *The Digital Dialectic*. Cambridge, Mass.: MIT Press, pp. 68–95.
Heeter, C. (1992) 'Being there: the subjective experience of presence'. *Presence*, 1 (2), 262–271.
Hegel, G. W. F. ((1977) *Phenomenology of Spirit*. trans. A. V. Miller. Oxford: Oxford University Press.
Heim, M. (1993) *The Metaphysics of Virtual Reality*. Oxford: Oxford University Press.
Henry, W. (2001) *Seminar on Personal Communications*. Ottawa: Carleton University, Department of Sociology and Anthropology.

Hermans, H.J.M. and Kempen, H.J.G. (1998) 'Moving cultures'. *American Psychologist*, 53 (10), 1111–1120.

Herz, J. C. (1997) *Joystick Nation: How Videogames Ate our Quarters, Won our Hearts and Rewired our Minds*. Boston, MA: Little, Brown & Co.

Hillis, K. (1999) *Digital Sensations: Space, Identity, and Embodiment in Virtual Reality*, Vol. 1. Minneapolis, London: University of Minnesota Press.

Hughes, E. (1993) 'A cypherpunk's manifesto'. In *Crypto Anarchy, Cyberstates and Pirate Utopias*. Cambridge, Mass.: MIT Press.

Hughes, J. A., O'Brien, J., Randall, D., Rouncefield, M. and Tolmie, P. (1998) *Some 'Real' Problems of 'Virtual' Organisation*. Working Paper, ESRC Virtual Society Research Programme (website). Retrieved from the World Wide Web: *http://www.brunel.ac.uk/research/virtsoc/projects.htm*.

Hum, P. (2001) 'The question after Sept. 11: Will they be viewed as privacy's champions or a security threat?' *The Citizen*, 6 December, pp. E4–6.

Jackson, P. J. (1999) 'Introduction: from new designs to new dynamics'. In P. J. Jackson (ed.), *Virtual Working. Social and Organisational Dynamics*. London: Routledge, pp. 1–18.

Jensen, M. (2000) 'Making the connection: Africa and the Internet'. *Current History*, 99 (637), 215–221.

Johnson, D. (1996) *Law and Borders: The Rise of Law in Cyberspace*. Retrieved July 2001 from the World Wide Web: *http://www.firstmonday.dk/issues/issue1/index.html*.

Jones, S. (1993) 'A sense of space: virtual reality, authenticity and the aural'. *Critical Studies in Mass Communication*, 10, n.p.

Kahn, J. (1999) 'The on-line brokerage battle'. *New York Times*, 4 October, C19.

Karleff, I. (1999) '"Electronic graffiti" leaves its mark'. *Financial Post*, 8 May, p. D8.

Katz, J. (2001) *Slashdot*. Retrieved August from the World Wide Web *www.slashdot.org*.

Keen, P. (1999) *Competing in Chapter 2 of Internet Business: Navigating in a New World*. Delft, The Netherlands: TU.

Keohane, R., and Nye, J. (2000) 'Globalization: what's new? what's not? (and so what?)'. *Foreign Policy*, 118, 104–130.

Kim, W. C. (2001) 'The future is a synthesis of bricks and clicks'. *Financial Times*, p. A10.

King, A.D. (1990) *Global Cities*. London: Routledge.

King, A.D. (1991) *Culture, Globalization and the World-system: Contemporary Conditions for the Representation of Identity*. Binghamton, NY: State University of New York at Binghamton, Department of Art and Art History.

Kingston, A. (2002) 'What's your excuse?' *National Post*, 16 March, pp. 1, 8.

Kirk, J. F. (2001) 'Forgotten office workers urged to take action'. *Toronto Star*, p. H7.

Kitchen, R. (1998) *Cyberspace: The World in the Wires*. Chichester, New York: John Wiley & Sons.

Kolakowski, L. (1985) *Bergson*. Oxford: Oxford University Press.

Krieger, M. H. (1986) 'Ethnicity and the frontier in Los Angeles'. *Society and Space*, 4, 385–389.

Kroker, A. and Weinstein, M. A. (1994) *Data Trash: The Theory of the Virtual Class*. Montreal: New World Perspectives.

Krueger, M. W. (1991) *Artificial Reality II*. Reading, Mass.: Addison-Wesley.

Laclau, E. (1996) *Emancipation(s)*. London: Verso.

Lanier, J. (1992) 'An insider's view of the future of virtual reality'. *Journal of Communication*, 42 (4), 150–171.

Lash, S. and Urry, J. (1994) *Economies of Signs and Space*. London: Sage.

Latour, B. (1993) *We Have Never Been Modern*. Cambridge, Mass.: Harvard University Press.

Latour, B. (2002) 'Is *re*modernization occurring – and if so, how to prove it?' *Theory, Culture and Society*, 19. Forthcoming.

Lefebvre, H. (1968) *Droit à la ville*. Paris: Anthropos.

Lefebvre, H. (1981) *La Production de L'Espace* (2nd edn). Paris: Anthropos.

Leonhardt, D. (2000) 'Computer technicians learn they are indispensable parts'. *New York Times*, 5 January.

Levy, P. (1998) *Becoming Virtual: Reality in the Digital Age* (trans. R. Bononno). New York: Plenum Press.

Lewicki, R. J. and Bunker, B. B. (1996) 'Developing and maintaining trust in work relationships'. In R. M. Kramer and T. R. Tyler (eds), *Trust in Organizations*. Thousand Oaks, Calif.: Sage, pp. 114–139.

Lewicki, R. J., McAllister, D. and Bies, R. (1998) 'Trust and distrust: new relationships and realities'. *Academy of Management Review*, 23 (1), 458–538.

Lipnack, J. and Stamps, R. (1997) *Virtual Teams: Reaching Across Space, Time and Organizations with Technology*. New York: John Wiley & Sons.

Lockard, J. (2000) 'Babel machines and electronic universalism'. In B. E. Kolko, L. Nakamura and G. B. Rodman (eds), *Race in Cyberspace*. New York: Routledge, pp. 171–190.

Ludlow, P. (ed.) (2001) *Crypto Anarchy, Cyberstates, and Pirate Utopias*. Cambridge, Mass.: MIT Press.

Luther, M. (1523) 'Vom Anbeten des Sakraments des heiligen Leichnams Christi'. In *Luther's Works*, Vol. 11. Vienna: Weimarer Ausgabe.

Lyotard, J-F. (1993) *The Interest of the Sublime* (trans. J. S. Librett). Albany, NY: SUNY Press.

McCarthy, C. (2000) 'The Barcelona Pavilion'. *Space and Culture – Virtual Space and Organizational Networks Issue*, 4/5, 87–98.

MacLennan, C. (1997) 'Democracy under the influence: cost–benefit analysis in the United States'. In J. Carrier (ed.), *Meanings of the Market*. Oxford: Berg, pp. 195–224.

McLuhan, M. (1964) *Understanding Media: The Extensions of Man*. New York: Routledge.

McNair, F. (2001) 'Climate ideal for Canadian telework book'. *Ottawa Citizen*, p. K4.

Maffesoli, M. (1996) *The Time of the Tribes*. London: Sage.

Maravall, J. A. (1986) *Culture of the Baroque: Analysis of a Historical Structure* (trans. T. Chochran). Manchester: Manchester University Press.

Matthews, J. (1997) 'Power shift: the rise of global civil society'. *Foreign Affairs*, 76 (1), 50-67.

Menzies, H. (1998) *Heather Menzies: New World Disorder*. Retrieved 18 February 2002 from the World Wide Web *http://www.mala.bc.ca/~soules/media212/new_wave/menzies.htm*.

Mieszkowski, K. (2001) 'Online Lolitas flirt with anonymous voyeurs'. *Ottawa Citizen*, 18 August, pp. B1, B8.

Miller, D. (1998) 'Conclusion: a theory of virtualism'. In J. Carrier and D. Miller (eds), *Virtualism*. London: Berg, pp. 187–216.

Miller, D. (2000) 'Virtualism – the culture of political economy'. In I. Cook, S. Naylor and J. R. Ryan (eds), *Cultural Turns/Geographical Turns: Perspectives on Cultural Geography*. Harlow, Essex: Pearson, pp. 196–213.

Mirchandani, K. (1999) 'Re-forming organisations. Contributions of teleworking employees'. In P. J. Jackson (ed.), *Virtual Working. Social and Organisational Dynamics*. London and New York: Routledge, pp. 61–75.

Mirman, M. J. and Bonian, V. G. (1992) '"Mouse Elbow": a new repetitive stress injury'. *Journal of the American Osteopath Association*, 92 (6), 701.

Mizroeff, N. (1999) *An Introduction to Visual Culture*. London and New York: Routledge.

Nandhakumar, J. (1999) 'Virtual teams and lost proximity. Consequences on trust relationships'. In P. J. Jackson (ed.), *Virtual Working. Social and Organisational Dynamics*. London and New York: Routledge, pp. 46–59.

Neill, M. (1995) 'Computers, thinking and schools in "The New World Economic Order"'. In *Resisting the Virtual Life*. San Francisco, Calif.: City Lights Books, pp. 181–194.

Neisser, U. (1982) *Memory Observed: Remembering in Natural Contexts*. San Francisco, Calif.: W. H. Freeman.

Nie, G. de (1998) 'Word, image and experience in the early medieval miracle story'. In P. Joret and A. Remael (eds), *Language and Beyond*. Amsterdam: Rodopi, pp. 97–122.

Novak, M. (1992) 'Liquid architectures in cyberspace'. In *Cyberspace, First Steps*. Cambridge, Mass.: MIT Press.

Nye, D. (1997) 'Shaping communication networks: telegraph, telephone, computer'. *Social Research*, 64, 1067–1091.

O'Connor, J. and Wynne, D. (1996) 'Left loafing: city cultures and post-modern lifestyles'. In J. O'Connor and D. Wynne (eds), *From the Margins to Centre: Cultural Production in the Post-industrial City*. Aldershot: Arena, pp. 49–90.

OECD (2001) *Understanding the Digital Divide*. Geneva: Organization for Economic Cooperation and Development.

OECD (2001) Draft Issues Paper: *Knowledge Management in the Public and Private Sectors: Similarities and Differences in the Challenges Created by the Knowledge-Intensive Economy*. Ottawa: Centre for Educational Research and Innovation, Organization for Economic Cooperation.

Oettermann, S. (1997) *The Panorama: History of a Mass Medium*, trans. D. L. Schneider. New York: Zone Books.

Osterman, A. L., Weinberg, P. and Miller, G. (1987) 'Joystick digit'. *Journal of the American Medical Association*, 257 (6), 782.

Pentland, B. (1997) 'Bleeding edge epistemology: practical problem solving in software support help lines'. In R. Barley and J. E. Orr (eds), *Between Craft and Science: Technical Work in US Settings*. Ithaca, NY: Cornell University Press, pp. 113–128.

Parker, P. (2001) 'Survey – mastering management: developing study for a not-so-global village'. *Financial Times* January 22, pp. 12, 14.

Perkins, T. (1999) 'The angler'. *Red Herring*, 14.

Pieterse, J. N. (1993) *Globalization as Hybridization*. The Hague: Institute of Social Studies Working Papers.

Pilkington, A. E. (1976) *Bergson and his Influence: A Reassessment*. Cambridge: Cambridge University Press.

Pimentel, K. (1993) *Virtual Reality, Through the New Looking Glass*. New York: Intel/Windrest/McGraw-Hill.

Pinsky, M. (1993) *The Carpal Tunnel Syndrome Book*. New York: Warner.

Podgorecki, A. (1992) *Nixon's Social Engineering Watergate as Social Engineering* (Working Paper). Ottawa: Department of Sociology and Anthropology, Carleton University.

Porter, T. (1986) *The Rise of Statistical Thinking 1820–1900*. Princeton, NJ: Princeton University Press.

Poster, M. (1990) *The Mode of Information: Poststructualism and Social Context*. Cambridge: Polity Press.

Poster, M. (2001) *What's the Matter with the Internet*. Minneapolis: Minnesota University Press.
Power, M. (1994) *The Adult Explosion*. London: Demos.
Purdy, J. (1998) *The God of the Digerati* (web page). Retrieved 25 April 2002 from the World Wide Web: *http://www.prospect.org/print/v9/37/purdy-j.html*.
Rawls, J. (1993) *Political Liberalism*. New York: Columbia University Press.
Robertson, R. (1992) *Globalization*. London: Sage.
Robertson, R. and Lechiner, F. (1985) 'The relativization of societies: modern religion and globalization'. In T. Robbins, W. Shepherd and J. McBride (eds), *Cultures, Culture and the Law*. Chicago, Il: Scholars Press.
Robinett, W. (1992) 'Synthetic experience'. *Presence*, 1 (2), 229–247.
Robinson, H. (ed.) (1846–1847) *Original Letters Relative to the English Reformation, Written During the Reigns of King Henry VIII, King Edward VI, and Queen Mary: Chiefly from the Archives of Zurich*. Cambridge: Parker Society.
Rodowick, D. N. (1999) 'The memory of resistance'. In I. Buchanan (ed.), *A Deleuzian Century?* Durham, NC: Duke University Press, pp. 37–57.
Ronell, A. (1989) *The Telephone Book: Technology, Schizophrenia, Electric Speech*. Lincoln and London: University of Nebraska Press.
Rorty, R. (1967) 'Relations, internal and external'. In P. Edwards (ed.), *Encyclopaedia of Philosophy*, Vol. 7. London: Collier-Macmillan, pp. 125–133.
Rose, G. (1984) *Dialectic of Nihilism*. Oxford: Blackwell.
Rothenberg, D. (1993) *Hand's End: Technology and the Limits of Nature*. Berkeley: University of California Press.
Rubin, A. M. (1988) 'A methodological examination of cultivation'. *Communication Research*, 15 (2), 107–134.
Rubin, J. (2001) 'What will the post-IT economy look like?' *Globe and Mail*, 8 September, p. 8.
Rucker, R. (ed.) (1992) *Mondo 2000 User's Guide to the New Edge*. New York: HarperCollins.
Salaman, G. (1997) 'Culturing production'. In P. Du Gay (ed.), *Production of Culture/Cultures of Production*. London: Sage, pp. 235–272.
Sambyal, A. and Kleiner, B. H. (2000) 'Developments concerning repetitive stress injuries'. *Management Research News*, 23 (78), 71–73.
Sardar, Z. (1996) 'alt.civilizations.faq:cyberspace as the darker side of the West'. In *Cyberfutures: Culture and Politics on the Information Superhighway*. London: Pluto Press, pp. 14–41.
Schachter, H. (2001) 'Make sense of data overload'. *Globe and Mail*, 8 August, p. M1.
Schachter, M. (2001) *Law of Internet Speech*. Durham, NC: Carolina Academic Press.

Schostak, J. (ed.) (1988) *Breaking into the Curriculum: The Impact of Information Technology on Schooling*. London: Methuen.
Scott, S. (2002) 'It's virtually a reality'. *National Post*, pp. 2, 6–7.
Shadid, A. (2001a) 'The legend and lessons of digital'. *Boston Globe*, 15 July, pp. E1, E5.
Shadid, A. (2001b) 'Preserving the past as path'. *Boston Globe*, 15 July, p. E5.
Shapiro, M. (1995) 'I'm not a real doctor, but I play one in virtual reality: implications of virtual reality for judgments about reality'. In D. McDonald (ed.), *Communication in the Age of Virtual Reality*. Hillsdale, NJ: Lawrence Erlbaum, pp. 323–345.
Shapiro, M. and Lang, A. (1991) 'Making relevision reality: unconscious processes in the construction of social reality'. *Communication Research*, 18 (5), 685–705.
Shapiro, S. P. (1987) 'The social control of personal trust'. *American Journal of Sociology*, 93 (4), 623–658.
Sheller, M. and Urry, J. (2000) *Automobilities*. Washington, DC: American Sociological Association Annual Conference.
Sheridan, T. B. (1992) 'Musings on telepresence and virtual presence'. *Presence*, 1 (1), 120–126.
Shields, R. (1989) 'Social spatialisation and the built environment: the case of the West Edmonton Mall'. *Environment and Planning D: Society and Space*, 7 (2), 147–164.
Shields, R. (1991) *Places on the Margin: Alternative Geographies of Modernity*. London: Routledge/Chapman and Hall.
Shields, R. (1993) *Life Style Shopping: The Subject of Consumption*. London: Routledge.
Shields, R. (ed.) (1996) *Cultures of Internet*. London: Sage.
Shields, R. (1997a) 'Flow'. *Space and Culture – Theme Issue on Flow*, 1, 1–5.
Shields, R. (1997b) 'Ethnography in the crowd: the body, sociality and globalization in Seoul'. *Focaal: Tijdschrift voor Antropologie*, 30/31, 23–38.
Shields, R. (2000) 'Hypertext links: the ethic of the index and its space-time effects'. In *The World Wide Web in Contemporary Theory: Metaphor Magic and Power*. New York: Routledge, pp. 144–160.
Shields, R. and West, K. (2000) *Innovation in Clean Room Construction: Cooperation Between Firms* (Working Paper). Ottawa: Innovation Management Research Unit, Carleton University.
Shove, E. (1996) *Factors Enabling and Inhibiting Innovation in the UK Construction Industry*. London and Lancaster: Lancaster University, Construction Sponsorship Directorate, UK Department of Environment.
Shulman, C. (2001) 'To the joystick born'. *Ottawa Citizen*, 12 February, pp. B2–3.

Silver, D. (2000) 'Margins in the wires'. In B. Kolko, L. Nakamura and G. Rodman (eds), *Race in Cyberspace*. New York: Routledge, pp. 133–150.

Simmel, G. (1990) *The Philosophy of Money* (trans. T. Bottomore and D. Frishy, 1st edn). London and New York: Routledge.

Simpson, L. (1995) *Technology, Time and the Conversations of Modernity*. New York: Routledge.

Slaughter, A.-M. (1997) 'The real new world order'. *Foreign Affairs*, 76 (5), 183-198.

Sobchack, V. (1995) 'Beating the meat/surviving the text'. In *Cyberspace/Cyberbodies/Cyberpunk*. London: Sage, pp. 205–214.

Sobchack, V. (1998) *Meta-morphing: Reflections on an Everyday and Yet Uncanny Phenomenon*. Retrieved 2000 from the World Wide Web *http://www.heise.de/tp/tpfhome.htm.*

Standage, T. (1998) *The Victorian Internet: The Remarkable Story of the Telegraph and the Nineteenth Century's Online Pioneers*. New York: Walker & Co.

Sterling, B. (1986) 'Introduction'.

Stevens, J. C., Witt, J. C., Smith, B. E. and Weaver, A. L. (2001) 'The frequency of Carpal Tunnel Syndrome in computer users at a medical facilities'. *Neurology*, 56 (11), 1568–1570.

Stewart, F. (1995) *Adjustment and Poverty: Options and Choices*. London: Routledge.

Stivale, C. (1998) *The Two-Fold Thought of Deleuze and Guattari*. New York: Guilford Press.

Stone, A. R. (1992) 'Will the real body please stand up?: Boundary stories about virtual cultures'. In M. Benedikt (ed.), *Cyberspace, First Steps*. Cambridge, Mass.: MIT Press.

Sutherland, I. (1965) *The Ultimate Display*. Washington, DC: ARPA.

Thomas, K. (2001) 'Hearing-impaired find a friend in IM'. *Burlington Free Press*, 24 July, p. 5.

Thompson, R. (2002) 'Drive-by hacking haunts wireless age'. *Financial Post*, 1 March.

Thrush, G. (2001) 'Robots make surgery "one big video game"'. *Ottawa Citizen*, 10 August, p. B1.

Tulloch, J. and Lupton, D. (2001) 'Risk, the mass media and personal biography: revisiting Beck's "knowledge, media and information society"'. *European Journal of Cultural Studies*, 4 (1), 5–28.

Turner, V. (1974) *Dramas Fields and Metaphors*. Ithaca, NY: Cornell University Press.

Tyler, T. R. (1984) 'The mass media and judgments of risk: distinguishing impact on personal and societal level judgments'. *Journal of Personality and Social Psychology*, 47 (4), 693–708.

United Nations Development Program (1999) *Human Development Report 1999*. Oxford: Oxford University Press.

Valente, T. W. (1995) 'Virtual diffusion or an uncertain reality'. In T. Bardini (ed.), *Communication in the Age of Virtual Reality*. Hillsdale, NJ: Lawrence Erlbaum.

Van Loon, J. (1997) 'Afterword'. *Space and Culture*, 1, 133–135.

Viera, P. (2002) 'Major banks not required to disclose credit scores to clients'. *National Post*, p. 1.

Wachtel, E. A. (1980) *Visions of Order*. Unpublished Ph.D. Dissertation, New York University.

Warschauer, M. (2000) 'Language identity and the Internet'. In B. Kolko, L. Nakamura and G. Rodman (eds), *Race in Cyberspace*. New York: Routledge, pp. 151–170.

Waters, M. (2001) *Globalization*. London: Routledge.

Weber, M. (1946) 'From Max Weber'. In H. H. Gerth and C. Wright Mills (eds), *From Max Weber: Essays in Sociology*. New York: Oxford University Press.

Weiser, M. and Seely Brown, J. (1996) *The Coming Age of Calm Technology*. Retrieved 15 February 2002 from the World Wide Web *http://www.ubiq.com/hypertext/weiser/acmfuture2endnote.htm*.

Whalley, P. (1997) 'Technical work in the division of labour: stalking the wily anomaly'. In S. R. Barley and J. E. Orr (eds), *Between Craft and Sciences: Technical Work in US Settings*. Ithaca, NY: Cornell University Press, pp. 23–52.

Widder, N. (2000) 'What's lacking in the lack: a comment on the virtual'. *Angelaki*, 5 (3), 117–138.

Williams, R. (1981) *Culture*. Harmondsworth: Penguin.

Williamson, O. E. (1993) 'Calculativeness, trust and economic organization'. *Journal of Law and Economics*, 30 (2), 131–145.

Winograd, T. and Flores, F. (1996) *Risky Business. Why People Feel Safe in Sexually Explicit On-line Communication*. Retrieved 25 February from the World Wide Web *http://www.ascusc.org/jcmc/vol2/issue4/witmer2.html*.

Woolley, B. (1993) *Virtual Worlds: A Journey in Hype and Hyperreality*. London: Penguin Books.

Wright, D. (2001) 'N.Y. cafe helps singles meet on Net, in person'. *Burlington Free Press*, 24 July, p. 4.

Wyatt, S., Henwood, F., Miller, N. and Senker, P. (eds) (2000) *Technology and In/Equality*. London: Routledge.

Zizek, S. (1989) *The Sublime Object of Ideology*. London: Verso.

INDEX